INTERIOR ELECTRICAL
WIRING

RESIDENTIAL

SIXTH EDITION

By

KENNARD C. GRAHAM

Electrical Engineer; Consultant, Apprenticeship Instruction
Material Laboratory of Bureau of Trade and Industry,
California State Department of Education

Authors of Previous Editions

ALBERT UHL

CARL H. DUNLAP

FRANK W. FLYNN

AMERICAN TECHNICAL SOCIETY · CHICAGO, ILL.

FIRST EDITION
1st Printing 1929

SECOND EDITION (REVISED)
2d Printing 1940

THIRD EDITION (REVISED)
3d Printing 1941
4th Printing 1944
5th Printing 1945
6th Printing 1946
7th Printing 1947
8th Printing 1949

FOURTH EDITION (REVISED)
9th Printing 1951

FIFTH EDITION (REVISED)
10th Printing 1952
11th Printing 1954
12th Printing 1956
13th Printing 1958
14th Printing 1960

SIXTH EDITION (REVISED)
15th Printing 1961
16th Printing 1963
17th Printing 1966
18th Printing 1968
19th Printing 1970
20th Printing 1970
21st Printing 1971
22nd Printing 1973

Published previously under the title
INTERIOR ELECTRIC WIRING AND ESTIMATING

PRINTED IN THE UNITED STATES OF AMERICA

PREFACE TO THE SIXTH EDITION

In an industry so characterized by the rapid development of new products and improved practices, it becomes periodically necessary to revise a book on this subject to adequately cover the latest developments. Therefore, the sixth edition of this popular book has been completely rewritten to meet the requirements of the electrician in the growing field of interior wiring.

The home of today is becoming a marvel of complexity with its use of modern products such as motors, speaker systems, remote control of lighting, heating and other functions. The successful electrician can no longer rest on what he has learned—he must keep abreast of what is new in the field in order to compete for his share of business in the increasing number of new homes being built.

It is the aim of this book to supply the information necessary to enable the electrician to adequately meet the increasing requirements of his field. Practical information on actual trade practices—a feature of each previous edition—has been retained and updated to fit the newer products and materials now on the market.

The emphasis throughout is on the advancement of the electrician's ability to do his job more skilfully and efficiently. It serves not only to acquaint the beginner with the field, but to keep the experienced worker abreast of new developments and techniques. The book progresses from basic theory through the different types of wiring to remote controls, motors, heating and farm wiring. Each chapter has been provided with new photographs and drawings to illustrate more fully the materials of the text.

The Publishers

CONTENTS

1. **Electrical Codes, Plans and Symbols**   1
   Purpose of regulations—Standards for Electrical Materials—Pre-
   liminary Considerations — Architect's Scales — Scaling — Common
   Scales—A Typical Plan—Electrical Outlets.

2. **Basic Electrical Theory**   9
   Fundamental Concepts — Three Primary Circuit Elements —
   Learning Ohms Law—Electrical Power and Energy—Series Cir-
   cuit—Parallel Circuit—Voltage Drop—Power Loss—Three-Phase
   Current—Network Systems.

3. **How to Make Electrical Connections** 22
   Approved Connections—Preparing the Wire—Soldering Copper
   Wire—Paste and Solder for Copper Wire—Making a Pigtail
   Splice—Making a Western Union Splice—Making a Tee Splice—
   Use of Rubber and Friction Tape—Use of Plastic Tape—Solder-
   less Splices — Splicing Small Copper Wires — Solderless Lugs —
   Splicing Aluminum Conductors.

4. **Installing Service and Metering Equipment** 33
   Electric Wiring Systems—Service Wiring—Service Drop—Service
   Wires — Underground Service Connections — Overhead Service
   Connections — Service Switch — System Grounds — Equipment
   Grounds — Grounding Conduit Systems — Grounding Armored
   Cable — Grounding Nonmetallic Sheathed Cable — Watt-hour
   Meter—Meter Wiring—Meter Installation—Bottom Connected
   Meter—Socket-type Meter.

5. **Lighting Outlets and Switches—Bell Wiring—Intercom**
   Systems 55
   Elementary Procedures—Reasons for Polarity Wiring—Single-
   Pole Switches—Two-Gang Switch—Three-Gang Switch—Double
   Pole Switches—Three-way Switches—Controlling Several Outlets
   —Four-way Switches—Operation of Controls at Three Points—
   Operation of Switches at Four Control Points — Electrolier
   Switches—Pilot-light Switches—Bell Wiring—Chime Circuits—
   Intercommunicating Telephones—Two-Station Arrangement—
   Four-Station Intercom System — Combination Loud-speaker/
   Intercom System.

6. **Methods of Wiring—Knob and Tube** 80
   Advantages of Knob-and-Tube Wiring—Open Knob-and-Tube
   Wiring—Insulators—Surface Clearance—Separation of Conductors
   —Protection on Ceiling, Side Walls and Floors—Support for
   Wires—Precautions Against Dampness and Acid Fumes—Stringing
   Wires—Concealed Knob-and-Tube Wiring—Essential Require-
   ments—Outlet Boxes—Method of Wiring New Buildings—Method
   of Wiring Old Buildings.

CONTENTS

7. **Methods of Wiring—Nonmetallic Sheathed Cable and Flexible Armored Cable** 94

Wiring Nonmetallic Sheathed Cable—Types—Applications—Essential Requirements—Method of Wiring New Buildings—Flexible Armored Cable—Types—Applications—Essential Requirements—Connectors—Polarity Grouping—Preparing the Cable—Service Entrance Cable—Feeder and Branch-Circuit Cable—Underground Feeder and Branch-Circuit Cable—Underplaster Extensions — Nonmetallic Surface Extensions — Surface Metal Raceways.

8. **Residential Wiring Processes** 106

New Houses — Sectional Drawings — Ceiling Finish — Floors — Mounting Heights—Construction—Boring Holes—Joists—Choosing Auger Bits—Sharpening Auger Bits—Old Houses—Preliminary Measures—Testing Circuits—Locating Outlets—Installing Switch Outlets—Plug Receptacle Outlets—Scuttle Holes—Fishing—Passing Obstructions—Removing Trim and Floor Boards—Removing Baseboards—Cutting Pockets—Replacing Wooden Floor—Installing Special Lighting Fixtures—Recessed Fixtures—Electrical Discharge Lighting.

9. **Methods of Wiring—Conduit** 126

Conduit Materials — Types of Conduit — Rigid Type — E.M.T. Type — Flexible Type — Conduit Elbows — Couplings — Bushings and Locknuts — Thin-wall Fittings — Connectors — Outlet Boxes — Fixture Studs and Hangers—Fastening Devices—Expansion Shells —Rawl Plugs and Drives—Homemade Anchorages—Wire Capacity of Conduit—Miscellaneous Shop Tools—Dies and Die Stocks—Hickeys and Pipe Benders—Right-Angle Bends in Conduit—Offsets in Conduit—Offsets and Saddles in Conduit—How to Install Thin-wall Conduit—Fishing.

10. **Large Appliances—Space Heating—Wiring for Motors** . . 152

Electric Ranges—Built-in Cooking Units—Dryers—Water Heaters —Central Heating—Duct Heating—Floor Furnaces—Wall Heaters and Portable Devices—Baseboard Heaters—Heated Ceilings —Ceiling Heaters—Thermostats—Line-voltage Control Units—Unit Heaters — Heat Pumps — Air Conditioning — Wiring for Motors—Connecting Single-Phase Motors—Connecting Three-Phase Motors.

11. **Multi-Family Dwellings; Special Construction Features** . . 174

Wiring Methods—Construction Procedures—Concrete Floors—Ground Slabs—Upper Floor Slabs—Flat Deck Slabs—Pan Slabs —Concrete Walls—Charges—Service Considerations—Switchboards and Panelboards—Stacking—Emergency Lighting.

12. **Residential Furnace Controls** 190

Automatic Control Equipment—Aquastat—Pressurestat—Airstat—Furnacestat—Motor-Operated Draft Control—Installing Heating-System Controls — Stoker Controls — Oil-burner Controls — Gas-burner Controls.

CONTENTS

13. **Remote Control Wiring**............................ 201

How Remote-Control Units Operate—Use of Remote Control Units in Wiring Systems—Advantages of Remote-Control Wiring —Equipment for Remote-Control Wiring—Transformers—Relays —Control Switches—Selector Switches—Wire—Planning Remote-Control Wiring—Remote-Control Layout—Locating the Relay— Outlet-Mounted Relays—Gang-Mounted Relays—Zone-Grouped Relays—Wiring Procedures—Installing Outlet Boxes—Installing Switches—Installing Relays—Installing Wire—Rewiring.

14. **Outside Wiring—Electrical Power on the Farm**.......... 215

Outside Wiring—NEC Rules for Outside Work—Farm Wiring —Buildings—Yard Distribution Plan—Yard Pole—Underground Wiring.

15. **Electrical Wiring Design**............................ 226

Basic NEC Requirements — Calculation of Load — Additional Loads—Optional Methods—Types of Occupancies Other Than Single-Family Dwellings.

16. **Estimating Electrical Wiring**........................ 233

Introduction — Preliminary Steps — Branch-Circuit and Fixture Schedules—Branch-Circuit Material Schedule—Other Methods of Take-off—Service and Feeder Material Schedules—Labor Unit Schedule — Estimating Forms — Final Considerations — Short-Cut Method for Estimating.

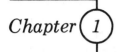

Chapter 1

Electrical Codes,
Plans and Symbols

QUESTIONS THIS CHAPTER WILL ANSWER

1. *Why is an electrical code needed to protect our homes?*
2. *What is meant by the letters NEC?*
3. *What are "scale" drawings?*
4. *How is switching indicated on a residential plan?*
5. *What symbols are used to show different types of outlets?*

Purpose of Regulations

When electricity was first applied to the lighting of homes several decades ago, one 25-watt lamp in each room was considered enough to supply average needs. The current taken by these few lamps amounted to not more than two, three, or perhaps five amperes. Number 14 AWG (American Wire Gage) rubber-covered copper wire, which had been generally accepted as the smallest size capable of withstanding the mechanical strain of installation, was able to carry this low current without overheating.

After the turn of the century, however, electrical consumption multiplied rapidly. Not only were stronger light bulbs introduced, but also such household appliances as irons and fans were made available to the general public. Installations previously considered adequate were then no longer able to carry the loads. As a result, overheating often destroyed wire insulation and caused fires, accidents, and even deaths.

Municipal authorities became interested; insurance companies and building industries began to take notice. Groups of interested parties clamored for laws to govern electrical wiring standards. Finally, the National Fire Protection Association was formed in Boston for the express purpose of preventing fires of an electrical origin. The set of regulations drawn up by this organization is known as the National Electrical Code, abbreviated NEC. This code is revised frequently, a new manual being issued every two years.

Although the Congress may not make the NEC a national electrical law since states are independent of the Federal government in regard to such matters, ordinances based upon this code, or taken from it word for word, have been enacted by states and cities. For this reason, the electrical worker should familiarize himself with NEC rules, and with such minor variations of these rules as may be recognized in his community. Throughout this text, NEC provisions are observed. In Canada, electrical construction work is governed by the Canadian Electrical Code, the contents of which are almost identical with those of NEC.

Standards for Electrical Materials

Even though electrical wiring is carefully installed, the materials and devices themselves can be sources of hazard unless properly designed for specific purposes. This fact caused the National Board of Fire Underwriters to create the Underwriters' Laboratories Inc. Manufacturers who wish to obtain U-L approval submit samples to one of the three testing laboratories—at New York, Chicago, or Santa Clara, California. Many large cities require U-L approval on all electrical materials used or sold within their jurisdiction.

Preliminary Considerations

Before starting electrical construction work, it is usually necessary to obtain a wiring permit from the city, town, or county having local jurisdiction. On large housing installations an architect delivers plans to his electrical engineer, who proceeds to mark in the outlets, conduit runs, panelboards, and other items. He also draws up specifications which set forth detailed standards for material and workmanship.

On smaller jobs, the engineer's services are seldom needed. The architect or draftsman furnishes the home builder with blueprints on which a comparatively small amount of electrical data is supplied. If electrical outlets are not indicated, their locations may be determined by a conference between homeowner, builder, and electrical contractor. The contractor is usually asked to sign a statement to the effect that he will use only approved materials, and that he will guarantee the completed installation to pass local inspection.

ARCHITECT'S SCALES

It is advisable for the student of interior electric wiring to obtain and familiarize himself with an architect's flat scale, such as Dietzgens'

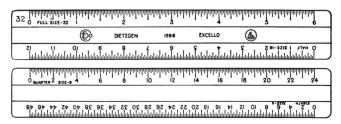

Fig. 1. Flat, or pocket scale

Courtesy of Eugene Dietzgen Co.

No. 1588, Fig. 1, for the following reasons:
1. Estimating wire lengths.
2. Determining locations for appliances.
3. Spotting fixtures and receptacles.
4. Drawing up circuit layouts that require minimum use of wire.

Scaling

Sometimes one or more required dimensions are missing from a blueprint. There may be several reasons for this; the main one is simplicity. To show every dimension would make the plans confusing. The process of obtaining the missing dimensions is called *scaling*.

Common Scales. The most common small scales are the $\frac{1}{8}$ inch equal to one foot which is written $\frac{1}{8}'' = 1'\text{-}0''$, and $\frac{1}{4}$ inch to one foot, or $\frac{1}{4}'' = 1'\text{-}0''$. Also used are $\frac{3}{8}'' = 1'\text{-}0''$; $\frac{1}{2}'' = 1'\text{-}0''$; $\frac{3}{4}'' = 1'\text{-}0''$; and $1'' = 1'\text{-}0''$. The scale that an architect selects depends upon the size of the building for which he is drawing the plans. For ordinary houses the $\frac{1}{4}'' = 1'\text{-}0''$ scale is suitable. For large buildings $\frac{1}{8}'' = 1'\text{-}0''$ is preferable in order to keep down the size of blueprints. In many cases the $\frac{3}{8}'' = 1'\text{-}0''$ scale is employed for elevation views because it is more suitable for complicated symbols, such as windows, than the $\frac{1}{4}''$ scale.

Architects often use what is called a three-sided, or triangular scale. This device, Fig. 2, is similar to a ruler except that instead of one edge, there are six edges, each of which has different graduations.

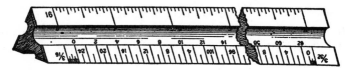

Fig. 2. Architects scale

In order to draw a line 25′ long according to the ⅛″ = 1′-0″, with the flat scale, Fig. 3, an architect actually draws it in 25 eighth-inch divisions. The eighths are numbered along the edge so they can be measured easily. Now suppose the line distance were 25′-6″. To draw this line, it is necessary to measure off 25 eighths plus one half

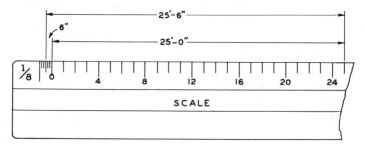

Fig. 3. Measuring with the flat scale

of another eighth-inch. A ⅛″ space at one end has 6 divisions, each of which represents two inches. Therefore, 3 of these lengths must be added to the 25 eighth-inches in order to mark off the whole distance.

A Typical Plan

The electrical drawing for a small frame bungalow is shown in Fig. 4. The house is 55′ long and 28′ wide. Although blueprints are drawn customarily to an "even" scale such as ⅛″ or ¼″ to a foot, as mentioned in the preceding section, the illustration here is drawn to an "uneven" scale in order to fit the page.

This one-story dwelling consists of a living room, dining room, kitchen, three bedrooms, bath, laundry, entry way, and central hall. Symbols for the most common types of electrical outlets are shown beneath the drawing. A ceiling outlet for an incandescent lamp is indicated by a small circle, a wall bracket by a circle with a supporting line or foot projecting from one side, a plug receptacle by a circle with two parallel lines drawn through it, and a switch by a capital *S*.

There are many different kinds of switches, the most common being single-pole, three-way, and four-way switches. A single-pole switch is represented by a plain *S*, a three-way by an *S* with the number *3* beside it, namely, S_3, and a four-way switch by the symbol S_4. Three-way switches enable one to turn a light on or off from widely separated points; for example, the two ends of a long hallway. A combination of two three-way switches and the required number of four-way switches permits a light to be operated from three or more points, should it be so desired. Methods for connecting these special switches will be explained later on.

In addition to electrical outlets, Fig. 4 shows doors, windows, partitions, closets, bathroom and kitchen fixtures. Swinging doors are indicated by a straight line and the arc of a circle, and sliding doors by short, straight lines which do not meet. Partitions are marked by parallel lines $\frac{1}{16}''$ or $\frac{1}{8}''$ apart for a 6″ partition, depending on whether the drawing is made to a $\frac{1}{8}''$ or $\frac{1}{4}''$ scale. Bathroom fixtures are shown by descriptive outlines: a bathtub, toilet, and lavatory here. A sink is located against the north wall of the kitchen.

Accordion doors are shown between kitchen and dining room, and between dining room and living room. Plaster arches are noted in two locations, between hall and living room, and between kitchen and laundry. A brick fireplace occupies a space along the east living room wall. Windows are shown in all four walls, those in the living room being single-pane, and all the others double-hung, with the exception of two casement windows in the dining room.

Electrical Outlets. Starting at the entry, a circle marks the overhead lighting outlet. A plug receptacle in the left wall can be used for connection to hedge trimmer, seasonal decorative lights, or other electric devices. A dashed line connects the outside light to another ceiling outlet inside the house. From this point, a dashed line con-

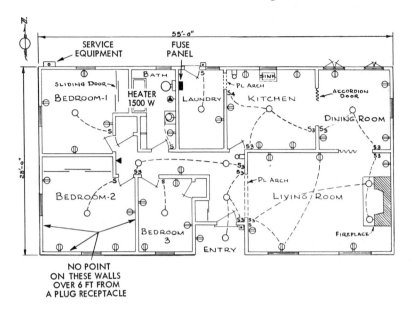

○	CEILING OUTLET	⊡	PUSH BUTTON
⊸○	WALL BRACKET OUTLET	⌁	BELL OR CHIMES
⊸⊖	PLUG RECEPTACLE (Convenience Outlet)	◄	TELEPHONE
⬤	SPECIAL PURPOSE OUTLET (Heater, For Example)	▬	FUSE PANEL

Fig. 4. Plan of bungalow

nects with a three-way switch to the right of the door. A second one connects to another three-way beyond the living room arch. These dashed lines are not intended to show paths of wires between outlets or switches, but merely to indicate which outlets are controlled by the various switches.

At this point, it should be mentioned that circuit wires, conduit, or cable runs, are not usually indicated in drawings for small, medium, or even large residences. To aid in explaining estimating procedure, however, circuit runs for this particular house are illustrated in Chapter 15.

Continuing with the survey, two ceiling lights, one in the entry and the other inside the front door, are controlled by a pair of three-way switches, as noted above. Another pair of three-way switches control two lights in the central hall, the first switch being located at the living room arch, the second on the wall between doors of bedrooms 2 and 3.

Entering the living room, bracket lights on either side of the fireplace are operated by two three-way switches, one at the arch on the west wall, the other at the side of the dining room door. A single-pole switch at the hall archway connects with two plug receptacles across the room from each other. Electric circuits will be explained later on. For the present, it may be noted that the NEC calls for a plug receptacle within 6 ft, horizontally, of every point of usable wall space in the rooms which it enumerates. Here, the 6-ft specification is observed with regard to plug receptacle outlets in living room, dining room, kitchen, and the three bedrooms.

The dining room ceiling light is controlled by two three-way switches at the sides of the accordion doors. The kitchen light is operated from three points. A three-way switch is located near the dining-room door, a second three-way is placed at the side of the hall door, and a four-way near the laundry arch. The laundry ceiling outlet is connected with a single-pole switch near the outside door. All plug receptacles in the kitchen and the laundry room are on special utility circuits, as required by NEC.

The bathroom has a bracket light, a plug receptacle, and a single-pole switch. There is also a 1500-watt wall heater, which is designated by a triangle inside a circle.

A fuse panel on the west wall of the laundry contains protective devices for circuits which radiate from it to the various outlets. This panel is supplied with current taken from service equipment, which is mounted on the north wall of the house adjacent to the west corner. The service installation includes a meter and a disconnect switch. Its details will be explained later.

Two bell pushes are shown, one near the entry door, the other beside the laundry door. A pair of two-tone chimes is provided, one in the central hallway, the other in the kitchen. One tone informs that the back, or north, push has been used, the other that the south button has been pressed. A telephone outlet occupies a space at the end of the central hallway, between bedrooms 1 and 2.

REVIEW QUESTIONS

1. What size of copper wire was first used in wiring homes?
2. Who writes the National Electrical Code?
3. Is it true that the NEC is seldom revised?
4. Can you name one of the early electrical appliances?
5. Who established the Underwriters' Laboratories, Inc.?
6. State the location of the Underwriters' Laboratory nearest your home.
7. What step is usually necessary before starting an electrical installation?
8. Who is responsible for seeing that the minimum permissible number of outlets is provided?
9. How are locations of outlets usually determined?
10. What device is usually employed in measuring plans?
11. What are the most commonly used dimensional scales?
12. What symbol represents an incandescent ceiling outlet?
13. What symbol represents a plug receptacle?
14. What symbol represents a four-way switch?
15. How are open archways indicated on the plan?
16. How can you tell, from looking at the plan, which switches control the various outlets?
17. What does NEC have to say about the number of plug receptacles in a bedroom, living room, dining room, or kitchen?
18. What symbol is used to represent the bathroom heater?
19. Are plug receptacles ever controlled by switches?
20. To what point does the power company deliver current for the house?

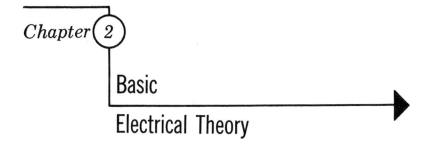

Chapter (2)

Basic

Electrical Theory

QUESTIONS THIS CHAPTER WILL ANSWER

1. What are the three expressions for Ohm's law?
2. What is the power equation?
3. How does a series circuit differ from a parallel one?
4. What is meant by the term line drop?
5. How does voltage drop result in power loss?

Fundamental Concepts

Current flows through an electrical conductor because of electrical pressure, in much the same way that water flows through a pipe because of water pressure. Figure 1 shows a simple electrical circuit

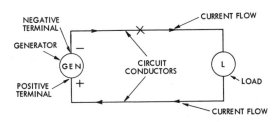

Fig. 1. Simple electrical circuit

which contains a source of electricity, lamp or other load which draws electricity, and connecting wires.

The source of electricity, which may be an electromagnetic unit, a battery, or other similar device, supplies the electrical pressure. This pressure sends an electrical current from the negative terminal of the generator through a connecting wire, the lamp, and the second wire, back to the positive terminal. The path formed by the conductors and the load is termed a *circuit*. Current will not flow unless the circuit is complete. Thus, if the wire is broken at point X, the flow will cease.

Three Primary Circuit Elements

The unit of electrical pressure is the *volt*. The unit of electrical current is the *ampere*. In order to find how many amperes will flow under a pressure of a given number of volts, it is necessary to know the resistance offered to passage of current by the material of which

Fig. 2. Voltmeter   Fig. 3. Ammeter
Courtesy of Weston Electrical Instruments Corp.

the conductors are made. The unit of electrical resistance is the *ohm*.

These units are so related that current in amperes is equal to pressure in volts divided by resistance in ohms. This relationship is known as *Ohm's law*, which may be written:

$$\text{Current} = \frac{\text{Pressure}}{\text{Resistance}} \text{ or Amperes} = \frac{\text{Volts}}{\text{Ohms}} \cdot$$

The following symbols are commonly employed to represent these three quantities:

$$\text{Pressure (volts)} = E$$
$$\text{Current (amperes)} = I$$
$$\text{Resistance (ohms)} = R$$

Ohm's law may be expressed, therefore, as:

$$I = \frac{E}{R} \cdot$$

If any two of the three quantities are known, the third may be determined by use of the formula, the unknown element being placed at the left of the equality sign, the other two at the right. Thus, if current and resistance are known, and it is desired to find what voltage is acting in the circuit, the formula may be written: $E = I \times R$. If current and voltage are known, the resistance of the circuit is found by writing the formula:

$$R = \frac{E}{I}.$$

A voltmeter for measuring voltage and an ammeter for measuring current are illustrated in Fig. 2 and Fig. 3.

Fig. 4. Ohms law

Fig. 5. Current

Fig. 6. Resistance

Fig. 7. Voltage

Learning Ohm's Law

Since Ohm's law embodies one of the fundamental principles of electricity, it is essential that it be memorized. A simple way of doing so is given in Figs. 4 to 7. If any one part is removed or covered, the relative position of the other two gives the value of the one covered in terms of the other two.

Thus, if I of Fig. 4 is covered, $E \div R$ is left, Fig. 5. Therefore, the value of I in terms of E and R is E divided by R. If R is covered, $E \div I$ remains, Fig. 6, giving the value of R in terms of E and I, which is E divided by I. In the same way, if E is covered, its value remains in terms of I and R, namely, I times R, Fig. 7.

A great amount of practice is required to learn how to apply

Ohm's law. Once the principle is firmly grasped, it is possible to handle a wide range of electrical problems. A few examples are given as follows:

Example 1

A voltage of 6 volts is used to force a current through a resistance of 3 ohms. What is the current?

Solution

The voltage (E) is 6 volts and the resistance (R) is 3 ohms. We wish to find the current (I). Using the first statement of Ohm's law we find that

$$I = \frac{E}{R} = \frac{6}{3} = 2 \text{ amperes}$$

Example 2

What voltage is required to force a current of 2 amperes through a resistance of 10 ohms?

Solution

The current (I) is 2 amperes and the resistance (R) is 10 ohms. We want to find the voltage (E).

$$E = I \times R = 2 \text{ amperes} \times 10 \text{ ohms} = 20 \text{ volts}$$

Example 3

A voltage of 20 volts is required to force a current of 5 amperes through a coil. What is the resistance of the coil?

Solution

Voltage (E) $= 20$ volts. Current (I) $= 5$ amperes

$$R = \frac{E}{I} = \frac{20 \text{ volts}}{5 \text{ amperes}} = 4 \text{ ohms}$$

Example 4

The voltage between the ends of a wire is 15 volts and its resistance is 3 ohms. What current will flow through it?

Solution

Covering the symbol I in the diagram, Fig. 5, there remains $E \div R$. Substituting the values of voltage and resistance given,
$$15 \div 3 = 5 \text{ amperes}$$

Example 5

A current of 10 amperes is forced through a conductor by a pressure or voltage of 30 volts. What is the resistance of the conductor?

Solution

Covering R in the diagram, Fig. 6, $E \div I$ remains. Substituting for E and I their values from the conditions as stated,
$$30 \div 10 = 3 \text{ ohms}$$

Example 6

A current of 10 amperes flows through a resistance of 2 ohms. What is the voltage that is forcing the current through the resistance?

Solution

Covering E, Fig. 7, we have left I times R. Substituting their values as before, we have $10 \times 2 = 20$ volts.

Electrical Power and Energy

The rate at which electric energy is delivered and consumed is called *power*. It is measured in *watts*. The *kilowatt*, which is 1000 watts, is used for the larger amounts of power. Thus, an electric-light bulb is rated at 120 volts and uses 50 watts of power. A larger bulb is rated at 120 volts and 200 watts. A 200-watt bulb requires four times as much power as the 50-watt bulb, even though both operate on the same voltage.

An electric generator supplying many lamps could be rated as 120 volts and 50 kilowatts. It is easier to read and write 50 kilowatts than to use 50,000 watts. A kilowatt is abbreviated *kw*.

When the amount of power in watts ends in three or more ciphers such as: 1000 watts, 2000 watts, 3000 watts, or 5000 watts, it is referred to as 1 kw, 2 kw, 3 kw, or 5 kw. If the power in watts is 1050 or 2780 watts or 5875 watts, the term watt is used instead of kilowatt.

The *power equation* is a rule for determining the amount of power in a circuit.

RULE: *Power equals the amperes flowing in a circuit times volts of the circuit.*

It can be written in simpler form by using the letters for abbreviation. Thus,

$$P \text{ is equal to } I \times E, \text{ or } P = IE$$

In the formula, P stands for power in watts. The letter I stands for current. It is the same letter used in Ohm's law. The letter E

Fig. 8. Power equation

Fig. 9. Power

Fig. 10. Current

Fig. 11. Voltage

stands for volts or voltage of the circuit. In formulas, the multiplication sign is often omitted as shown at the right. The power equation can be arranged as shown in Fig. 8. Here a square is used instead of a circle so it would not be confused with Ohm's law. The method of applying it is the same.

To find the power of a circuit, place a finger on the letter P, as shown in Fig. 9. Then $I \times E$ gives power. It is read, *current in amperes times voltage gives power.*

In like manner, to find the current represented by the letter I, place a finger on that letter, Fig. 10. Then, $I = \dfrac{P}{E}$, which is the same as $I = P \div E$. Thus, current equals power in watts divided by voltage.

When power and current are known, the voltage rating is found by placing a finger on the letter E, Fig. 11, so that $E = \dfrac{P}{I}$, which means E is equal to $P \div I$.

Following are several different voltages and currents which will produce the same amount of power in watts:

$$6 \text{ volts} \times 200 \text{ amperes} = 1200 \text{ watts}$$
$$12 \text{ volts} \times 100 \text{ amperes} = 1200 \text{ watts}$$
$$24 \text{ volts} \times 50 \text{ amperes} = 1200 \text{ watts}$$
$$120 \text{ volts} \times 10 \text{ amperes} = 1200 \text{ watts}$$
$$240 \text{ volts} \times 5 \text{ amperes} = 1200 \text{ watts}$$
$$600 \text{ volts} \times 2 \text{ amperes} = 1200 \text{ watts}$$

As the voltage increases, the amount of current required for the same power is reduced. In an automobile, a 6-volt storage battery, located a few feet from the engine, may supply 200 amperes. In residential wiring where the distance from the power company's transformer to the customer's home is a hundred feet or more, the voltage is 115 for a two-wire service, or 115-230 for a three-wire service. The current flowing to the residence, at 115 volts, would be only slightly more than 10 amps for a 1200 watt load. Voltage loss or voltage drop is not desirable because it reduces voltage to the lamps or other appliances in the circuit.

Series Circuit

There are two common types of circuits: *series* and *parallel*. The series form is illustrated in Fig. 12. Part (A) shows a string of Christmas tree lights which are connected in series. Part (B) is a line diagram of the circuit. Current flows through each lamp, one after the other, before returning to the supply wires or to the generator.

Certain facts should be observed in regard to the series circuit:

1. The current flowing in all parts of a series circuit is the same. This fact can be proved by connecting an ammeter anywhere in the circuit and noting that it reads the same at all points.

2. Removing any lamp or electrical appliance from a series circuit interrupts the flow of current.

3. The amount of current is inversely proportional to the resistance of the units connected in that circuit (as units of greater resistance are substituted, the current will be reduced).

4. The total voltage is involved in doing work in that circuit. If a voltmeter with insulated flexible leads is used to take successive

readings around the circuit, the total of those readings will equal the applied voltage.

If ammeter No. *1* reads 1 ampere, ammeter No. *2* will also read 1 ampere because all of the current (amperes) leaving the generator returns to it. This is an important point to remember in all circuit work no matter how complicated.

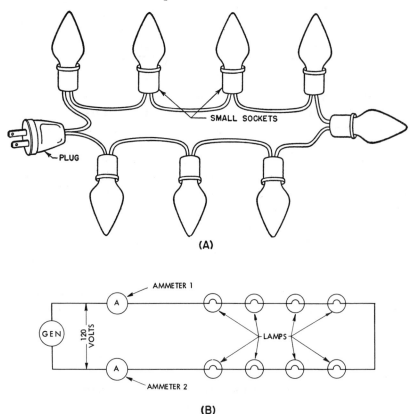

Fig. 12. Series circuit

Note that the ammeters in Fig. 12 are part of the series circuit. An ammeter is always connected in this manner, so that it is necessary to interrupt the flow of current, at least momentarily, while the instrument is "cut into" one or the other of the circuit conductors.

Parallel Circuit

In the parallel arrangement, Fig. 13, each lamp is connected in a sub-circuit of its own, a complete path being formed through the generator and connecting wires without involving the other lamps. Important facts associated with the parallel circuit are:

1. The current flowing in the main lines of a parallel circuit is the sum of the currents in the various parallel paths of that circuit. If each lamp in Fig. 13 takes one ampere, the total current will be 5 amperes, as indicated.

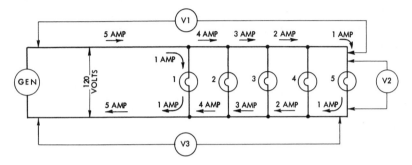

Fig. 13. Parallel circuit

2. Removing any lamp or electrical appliance from a parallel circuit does not break the flow of current to other units. For example, removing lamp L_1 would not prohibit the other lamps from operating.

3. The amount of current flowing in a parallel circuit is directly proportional to the number of units connected. (As more lamps are connected, more current flows through the wires.)

4. The total voltage is involved in doing work around any closed path from the generator. If a voltmeter with insulated flexible leads is used for successive readings around any closed loop, the total of the voltage drops will equal the applied voltage.

If three voltmeters are used, as shown in Fig. 13, the reading of voltmeter No. *1* will equal the voltage lost in the section of line across which it is connected, namely, 2 volts. In like manner, voltmeter No. *2* will read the voltage lost in lamp No. *5*, 116 volts. Voltmeter No. *3* will read the voltage lost in the return line to the generators, 2 volts. Adding the voltage drops, 2 + 116 + 2 equals 120 volts, which is generator voltage.

The voltage lost in the two wires feeding lamp *5* depends on the length of the line wire.

If this same test is made around any other closed circuit, it will be found that the total of the voltage losses equals the generator voltage.

Note that the voltmeter is connected in parallel with all or part of the circuit, as desired. Thus, the current is not interrupted when using the voltmeter to take readings.

Voltage Drop

When current flows in a conductor, part of the voltage is lost in overcoming resistance. If the loss is excessive, that is, more than a few percent, the lamps or other devices will not operate satisfactorily. Lamps decrease in brilliancy, heating devices deliver a reduced output, and motors have difficulty in starting their loads.

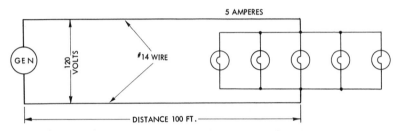

Fig. 14. Generator supplying lighting load

According to Ohm's law, voltage $E = I \times R$. This formula may be used to determine voltage drop. If the resistance of the circuit wires is 0.5 ohm, and the current 20 amperes (hereafter abbreviated "amps"), the voltage drop will be 20 amps × 0.5 ohm, or 10 volts.

In most cases, the resistance of the wires is not given, but their size is known. A simple formula can be employed to find resistance. For example, in Fig. 14, a circuit of two No. 14 AWG copper wires supplies 5 lamps, each of which consumes 1 amp, at a distance of 100 ft from the source of supply. It is desired to calculate the voltage drop in the circuit wires. In order to find voltage, current in amps and resistance in ohms must be known. Here, only the current is known.

The resistance formula reads as follows: $R = \dfrac{K L}{D^2}$.

R = resistance in ohms (when found)

K = a constant whose value depends on the conductor; for example, 10.8 for copper wire, or 17 for aluminum wire

L = length in feet (both wires)

D = diameter of the wire in thousandths of an inch or mils (1/1000-inch is known as a "mil")

D^2 = diameter squared (multiplied by itself), this diameter being expressed in mils, and the squared result in "circular mils." Electrical tables give the circular mil areas of wires.

The length, here is 2 × 100 ft, or 200 ft.

The diameter of No. 14 AWG wire is 64.1 mils.

The circular mil area (diameter squared) is equal to 4107.

Substituting these values in the formula:

$$R = \frac{10.8 \times 200}{4107} = \frac{2160}{4107} = .526 \text{ ohm.}$$

Voltage drop equals:

$E = I \times R = 5$ amps × .526 ohm = 2.63 volts.

If the voltage at the source is 120, that at the lamps must be equal to: 120 volts − 2.63 volts, or 117.37 volts.

This drop would not be considered too far out of line, but if the current were 10 amps, it would amount to twice 2.63 volts, or 5.26 volts, which would be somewhat high. In order to cut down the loss of voltage, wires of a larger diameter must be employed.

The next larger size is No. 12 AWG, which has a diameter of 81 mils, and a circular mil area of 6530. Substituting this value in the formula:

$$R = \frac{10.8 \times 200}{6530} = \frac{2160}{6530} = .331 \text{ ohm.}$$

Voltage drop now equals: 10 amp × .331 ohm = 3.31 volts. This drop is still high, so that a larger diameter wire must be used. The next larger size is No. 10 AWG, with a diameter of about 102 mils and a circular mil area of 10,380. The resistance of 200 ft of this wire is equal to: 2160/10,380, or .208 ohm. Voltage drop becomes: 10 amps × .208 ohm, or 2.08 volts, which is acceptable.

Power Loss

Voltage drop results in power loss, which simply heats up the wires. This power loss is registered on the meter, and is paid for by the customer. According to the power formula, Power = $E \times I$. Power Loss = E (line drop) × I.

If the current in a long feeder wire is 50 amps and the resistance is .35 ohm, the voltage drop = 50 amps × .35 ohm, or 17.5 volts. The power loss = 17.5 volts × 50 amps, or 875 watts, almost enough to keep nine 100-watt lamps burning continuously.

Three-Phase Current

Only single-phase current has been considered up to this point. As will be seen later, motor loads are frequently supplied by three-phase current. A three-phase supply, Fig. 15, consists of three wires, the phase voltages (that between any two wires) being exactly alike. Here, the voltage is 230 volts. Drop in any one of the supply wires

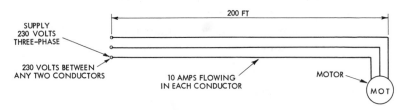

Fig. 15. Three-wire, three-phase system

is determined in the same way as for single-phase current. In order to determine the drop per phase, this value is multiplied by the number 1.73.

For example, the three No. 10 conductors in Fig. 14 carry a 10-amp, three-phase current to a motor which is 200 ft from the supply point. The voltage drop in a single wire 200 ft long has already been calculated above as 2.08 volts. The voltage drop per phase equals 1.73 × 2.08 volts, or approximately 3.6 volts. The voltage delivered at motor terminals, therefore, is 230 volts − 3.6 volts, or 226.4 volts.

Network System

Both lighting and power loads may be supplied by the network system of three-phase distribution illustrated in Fig. 16. Three-phase motors are connected to the main or phase wires, lighting circuits between any one of the main wires and the neutral conductor. The voltage between any two of the phase wires is 208 volts, while that between any phase wire and the neutral conductor is 120 volts. In a perfectly balanced system, where only lighting circuits are supplied, voltage drop per circuit and power loss per circuit are equal only to that of a single conductor. Total loss in all three phases, under this condition, is equal to three times that in a single conductor.

Using values given in the above calculation, where the distance from load to center of distribution is 200 ft, the current 10 amps, and the size of conductor No. 10, the voltage drop per circuit is equal to 2.08 volts, and the voltage supplied to the lamps equals 120 volts minus 2.08 volts, or approximately 118 volts. The power loss in each wire is $I \times E$, which is 10 amps $\times$ 2.08 volts, or 20.8 watts. The power loss in all three wires is three times this amount, or 62.4 watts.

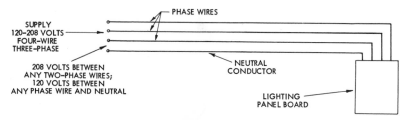

Fig. 16. Four-wire, 120-208 volts, three-phase, network system

REVIEW QUESTIONS

1. What term describes the path formed by the conductors and the load.

2. Name the three primary circuit elements.

3. State Ohm's law when voltage and resistance are known.

4. State Ohm's law when voltage and current are known.

5. State Ohm's law when current and resistance are known.

6. What current flows when the circuit voltage is 120 and its resistance is 10 ohms?

7. What circuit voltage is required to force 5 amps through a 20-ohms resistance?

8. How many amps will flow if circuit voltage is 100 and resistance is 40 ohms?

9. What happens to the circuit if one of ten *series* lamps burns out?

10. What happens to the circuit if one of ten *parallel* lamps burns out?

11. Is the ammeter connected in parallel or in series with the circuit?

12. Is the voltmeter connected in parallel or in series with the circuit?

13. What effect does resistance have on voltage delivered to the load?

14. What is meant by the term *power loss*?

15. State the power equation.

16. State the formula for determining resistance of a given size of copper wire.

17. Would the resistance increase or decrease if aluminum were substituted for copper in the wire?

18. What is the voltage drop when resistance is 10 ohms and the current 12.5 amps?

19. What is the resistance of 100 ft of No. 14 copper wire if its circular mil area is 4107?

20. What power loss results if 10 amps flow through the above wire?

Chapter ③

How to Make
Electrical Connections

QUESTIONS THIS CHAPTER WILL ANSWER

1. What is the best method for removing insulation from wires?
2. How are copper joints soldered?
3. What is the procedure in making a tee splice?
4. How are wires joined with solderless connectors?
5. How are aluminum conductors joined?

Approved Connections

The importance of making good splices and taps in the wiring systems cannot be overemphasized. No connection will be as good mechanically as an unspliced wire, but approved methods of splicing, taping, soldering, or installing solderless connectors will produce a satisfactory joint. On the other hand, connections made in outlet boxes and at switches or receptacles by an inexperienced workman are often the weakest parts of an interior wiring system.

Preparing the Wire

The first step in making any kind of wire connection is to remove the insulation, a process commonly known as *skinning*. For this purpose, use a fairly sharp knife. Do not circle the wire with the blade at right angles, because in most cases this produces a groove in the conductor which may cause the latter to break when bent at this point. Instead, whittle the insulation away in a manner similar to that of sharpening a pencil, as in Fig. 1.

High-grade rubber and plastic insulations are tough and do

not cut easily. Draw the knife across the material instead of straight along it. The higher the quality of insulation, the sharper the knife must be. Insulation may be removed from ends of small wires by means of a tool called a *wire stripper,* Fig. 2.

Removing Weatherproof Insulation. Insulation on weatherproof wire, which is used for outdoor work, is tough and hard to skin, especially after becoming thoroughly weathered. Electricians lay the end of the wire on the pavement and pound with a hammer to crack

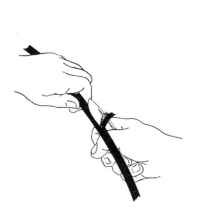

Fig. 1. Removing insulation

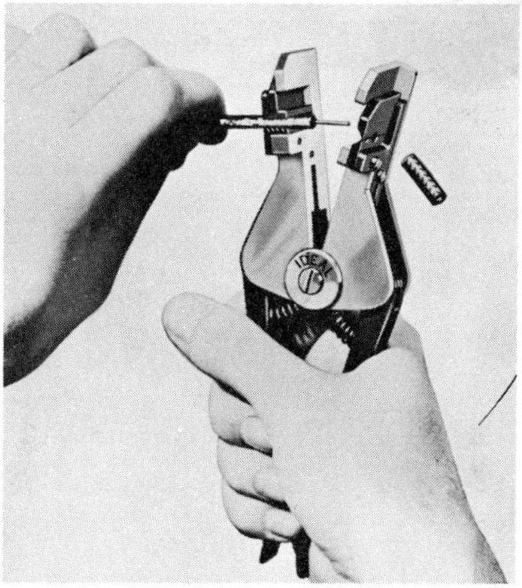

Fig. 2. Wire stripper
Courtesy of Ideal Industries, Inc.

the braid and loosen it from the conductor. There is nothing wrong with this method when care is exercised so that the face of the hammer strikes the wire squarely and yet not hard enough to deform the conductor. The asphalt, with which the braid is impregnated, does not readily come away from the copper even after the braid has been removed; however, a rag and some gasoline will help. White gasoline, which does not contain lead, is preferable.

Soldering Copper Wire

Soldering can be done with the aid of a soldering iron, a soldering gun or a torch, Fig. 3. A common method is to use a methane or propane torch. The joint is first cleaned, covered with a thin layer of soldering paste, heated in the flame of a torch, and then treated with a piece of string solder. The solder melts upon contact, metal flowing readily along and between the conductors.

Paste and Solder for Copper Wire

Soldering paste should be non-corrosive. Ordinary commercial pastes, Fig. 4A, satisfy this requirement. Solder commonly used by electricians is 50/50 grade—50% tin and 50% lead. Various combinations of these metals are obtainable, such as a 40/60 mixture, but the 50/50 alloy, Fig. 4B, flows more readily, and does not call for so much heat as do some other grades.

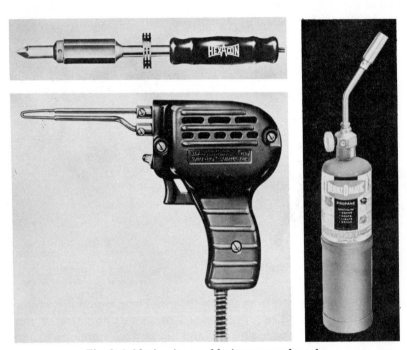

Fig. 3. Soldering iron, soldering gun and torch
Courtesy of Hexacon Electric Co.; Wen Products, Inc.; Otto Bernz Co., Inc.

SOLDERED SPLICES, COPPER WIRES

Three Common Types

Splices are of three different kinds: (1) the *pigtail,* used for connecting the wire ends in outlet boxes; (2) the so-called *Western Union splice,* used principally on solid conductors which must carry their own weight for a considerable span or of a size too large for a pigtail; (3) the *tee tap,* used to connect one wire to a continuous run of another wire.

Making a Pigtail Splice. Skin the wires an inch or so, as in Fig. 5A. Cross the ends as in Fig. 5B. Hold them in position with thumb and fingers of one hand while twisting them together with pliers for six or eight turns, as in Fig. 5C. Cut off the projecting wires ½" beyond the last twist, as in Fig. 5D. Then, double the end back with the pliers, as in Fig. 5E, to eliminate a sharp point

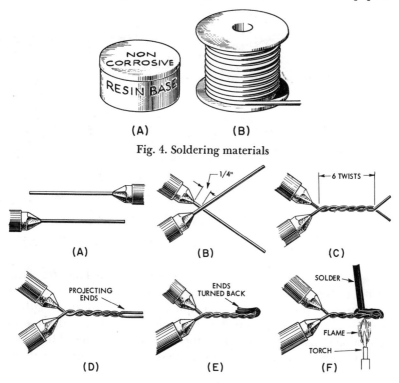

(A)   (B)

Fig. 4. Soldering materials

(A)   (B)   (C)

(D)   (E)   (F)

Fig. 5. How to make a pigtail splice

that may puncture insulation. Solder the splice as in Fig. 5F. Two, three, or four wires can be joined readily in this manner.

Making a Western Union Splice. Skin insulation from the wires a distance which depends on their size, varying from about 3" with No. 14 AWG to perhaps 6" for No. 10. Cross the wires at their middle as in Fig. 6A. Twist the ends in opposite directions four to six turns, Fig. 6B. Now twist each end sharply, at right angles to the run of the splice, and wind three full turns, Fig. 6C. Cut off the excess ends, and solder, Fig. 6D.

Making a Tee Splice with No. 14 Wires. Remove the insulation on the main wire for a distance of 1¼″, and that on the tap wire 2″, as in Fig. 7A. Cross the wires as in Fig. 7B. Twist the tap wire around the main wire, Fig. 7C, and draw the turns tight by using pliers, Fig. 7D. Cut off excess wire and solder, Fig. 7E.

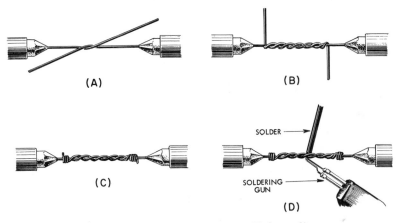

Fig. 6. How to make a Western Union splice

Taping Splices—Two Methods

There are, in general, two widely accepted methods for insulating a soldered splice, the purpose in either case being to cover the joint with protective material equivalent in electrical strength to that on the original conductor. One method calls for the use of rubber and friction tape; the other, for plastic tape. The older method, which is applicable to a Western Union splice, will be explained first.

Use of Rubber and Friction Tape. Cut a 4″ piece of rubber from the roll. There is a strip of glazed cotton tape on one side of the rubber to prevent adjacent layers from sticking together. Remove this material, and stretch the rubber to a length of about 10″.

Starting at the left-hand end of the splice, Fig. 8A, make a turn around the wire to secure the tape. Now, wrap the joint diagonally toward the right, keeping the tape stretched and overlapping half the width of the preceding turn. At the right-hand end, the direction of travel is reversed, as in Fig. 8B, the wrapping continuing until two layers have been applied. Excess tape is cut off, and the end pinched against the joint with thumb and forefinger, Fig. 8C, the heat of the fingers causing the rubber to stick to itself. A layer of friction tape is now applied in the same manner, Fig. 8D. Since friction tape tends to unwind as it ages, a quick-drying insulating paint is often added to prevent such occurrence.

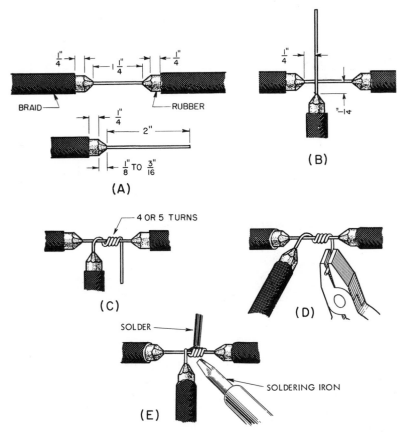

Fig. 7. How to make a tee splice

Use of Plastic Tape. Plastic tape has come into general use in recent years because of the saving in labor and in space, over the rubber-friction method. Figure 9 shows the taping of a pigtail splice with this material. As may be seen in Fig. 9A and Fig. 9B, the procedure is much the same as explained above. Although the manufacturer states that a single layer is sufficient for voltages up to

600 volts, it is well to apply two layers of half-lap tape for added mechanical protection.

Solderless Splices

Splicing Small Copper Wires. Wire nuts were first used for splicing fixture wires to circuit conductors. A wire nut, illustrated in Fig. 10, consists of a cone-shaped spiral spring inside a molded

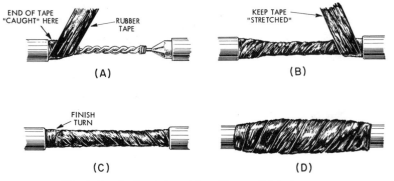

Fig. 8. Insulating joint with rubber and friction tape

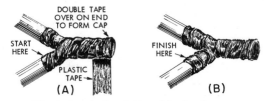

Fig. 9. Insulating joint with plastic tape

body. When screwed onto the wires, the spring presses them firmly together, and the insulating body protects the wires from external contact, so that no tape is required.

Sherman fixture connectors, Fig. 11, make use of set-screws to hold the wires in place. With this device taping is, of course, necessary.

Wire nuts are now made which can be used to connect circuit wires up to No. 8 AWG, the Scotchlok device of Fig. 12 being a good example. But the crimp type of solderless connector, Fig. 13, is quite popular. The wires are first twisted together and a copper thimble slipped onto the joint. A special, plier-like tool is used to crimp thimble and wires together. The joint may then be taped, or fitted with a vinyl insulating cap, as shown in the illustration.

Fig. 10. Wire nut
Courtesy of Ideal
Industries Corp.

Fig. 11. Sherman fixture connector

Courtesy of H. B. Sherman Manufacturing Co.

Fig. 12. Scotchlok connectors
Courtesy of Minnesota Mining and Manufacturing Co.

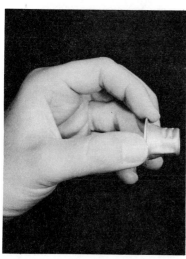

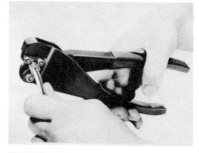

Fig. 13. Installing crimp type solderless connector
Courtesy of Buchanan Electric Products Corp.

The solderless terminals of Fig. 14 have largely replaced small solder lugs. One type, as seen in the illustration, is fastened by crimping, or indenting by means of a special tool.

Fig. 14. Solderless terminals

Courtesy of Thomas and Betts Co.

Solderless Lugs. There are three general types of solderless lugs: *set-screw, compression,* and *indented,*Fig. 15. The latter are installed by means of hydraulic pressure devices, or by tools which employ compound leverage. Welding is also used to some extent in making solderless connections.

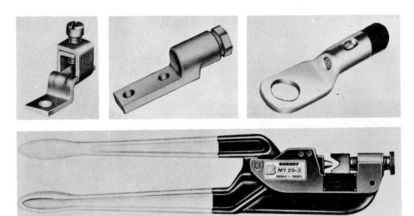

Fig. 15. Connectors and indenter for large conductors

Courtesy of Thomas and Betts Co., Penn-Union Electric Co., Burndy Corp.

Splicing Aluminum Conductors. Aluminum conductors can be soldered with special solder and flux, but they are usually joined by other methods involving pressure, compression, or welding, as in Fig. 16. Experience has shown that the best results are obtained when aluminum fittings are used with aluminum conductors. Recently, however, plated copper alloy fittings have proved suitable for use with aluminum. In any case, a special compound should be

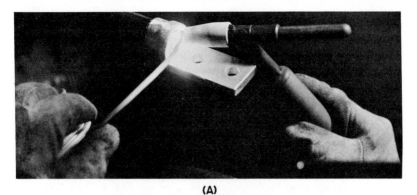

(A)

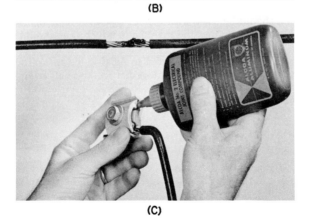

(B)

(C)

Fig. 16. Terminating aluminum conductors
Courtesy of Aluminum Company of America

applied to surfaces of conductors and lugs at points of contact in order to prevent corrosion and to ensure good electrical contact. When copper and aluminum conductors are connected together, the splicing device must have a separate slot or compartment for each type of conductor. Attempts to hold them together under the same screw or compression nut will lead to ultimate trouble on account of corrosive galvanic action.

REVIEW QUESTIONS

1. State the most likely weak point of an interior wiring installation.
2. Would you remove weatherproof insulation from copper wire by paring?
3. What does the expression 50/50 mean, with respect to solder?
4. How can you eliminate the sharp end on a pigtail splice?
5. Is the Western Union splice used in making a tee joint?
6. How are wire nuts installed?
7. Is rubber tape used in conjunction with plastic insulating material?
8. How are soldering tongs heated?
9. What is the most common method of heating splices?
10. What is a compression connector?
11. What is a crimp connector?
12. Is the wedge-on connector of the crimp type?
13. Can aluminum wires be soldered?
14. Can copper and aluminum wires be soldered together?
15. Why is a joint compound used on aluminum terminals?
16. Must aluminum terminals always be used with aluminum conductors?
17. What should be done with rubber tape as it is being applied?
18. What term describes the manner in which tape should be applied to a joint?
19. Is the use of wire nuts limited to making fixture connections?
20. What device holds the wires together in the small crimp connector?

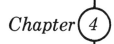

Chapter 4

Installing Service

and Metering Equipment

QUESTIONS THIS CHAPTER WILL ANSWER

1. What is the difference between service drops and service conductors?
2. How is an overhead service installed?
3. Why must service equipment be bonded?
4. What points should be observed in making ground connections?
5. How are watt-hour meters connected?

ELECTRIC WIRING SYSTEMS

The electric wiring system in a residential building may be divided into separate sections or classifications, each of which has its specific purpose. In Fig. 1, relationships between the various sections are shown, and the manner in which they are connected.

The main divisions of a complete wiring installation may be listed as follows: service drop, service wires, service switch, grounding, and metering. In succeeding paragraphs each of these parts is defined, and the function of each member is explained in detail.

SERVICE WIRING

Service Drop

The lead-in wires from the power company lines to the secondary rack on the customer's building are known collectively as the *service drop*.

The service drop, shown in Fig. 1, is actually not part of the wiring system in a building. It is the connection provided by the power company from area distribution lines to the building. It is

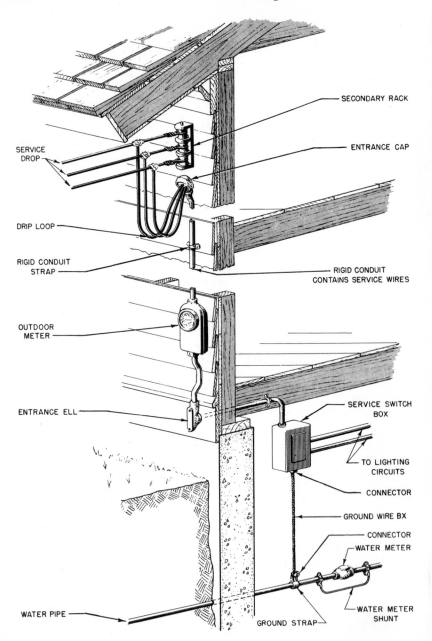

Fig. 1. Principal parts of a wiring system

discussed here because failure to observe important points relating to the service drop may result in delay and expense.

Service Wires

Service wires connect the power company service drop to the service switch.

The wires may be carried to the building either underground or overhead, depending upon the location of the power company distribution lines. The electrical wireman should know: (1) whether

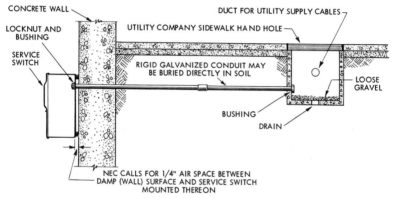

Fig. 2. Underground service conduit to sidewalk handhole

underground or overhead service will be used; (2) the voltage and amperage; (3) whether single- or three-phase; (4) whether two, three, or four wires will be employed; (5) the type of watt-hour meter, and where it will be located; (6) and the point at which these wires are to enter the building.

Underground Service Connections

These connections are of three general types: (1) from the service switch directly to a utility company manhole in the street; (2) from the service switch to a utility company sidewalk handhole; and (3) from the service switch to a pole riser.

When the service originates in the manhole, it is necessary to make arrangements with the company for joint installation of conduit and wire because no one except a utility company workman is permitted to enter the chamber.

When the service originates in the sidewalk handhole, Fig. 2, the electrical contractor is usually permitted to run his conduit and wire into it. Utility personnel connect the wires to service cables which pass through ducts from one handhole to another along the street.

A pole riser service, Fig. 3, is also a joint enterprise between electrical contractor and the utility company. In many cases, the electrical contractor simply installs conduit to a point at least 8' above grade, the minimum height for such protection as specified by the NEC, and pulls in enough wire to reach the crossarm at the top of the pole.

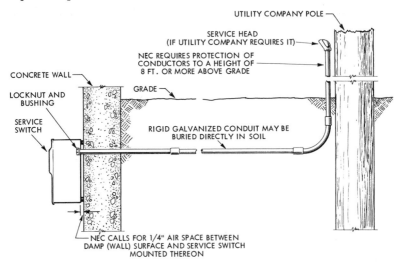

Fig. 3. Underground service conduit and pole riser

Utility linemen extend the wires up the pole, covering them with approved wooden molding. In some cases, a short piece of conduit and a service head are used to terminate the run at the crossarm, the contractor supplying these items to the company. In other cases, the contractor must furnish a service head and enough conduit to reach the desired point.

In new construction, entry through the wall is made before the concrete is poured, a single length of conduit being secured in place as forms are assembled. A few years ago, lead-covered wires had to be used for underground service. With the advent of newer types of insulations, however, lead has in most cases been replaced by type RW-RH or RHW conductors.

Overhead Service Connection

The overhead *service drop* is usually installed by the power company. It is secured to the building by appropriate means at a point where the line gang can obtain a direct pull from the nearest service pole. The wires must be high enough to provide proper clearance above grade, and must not come within 3′ of any door, window opening, or fire exit. The exact location of the pull-off should be learned from the power company.

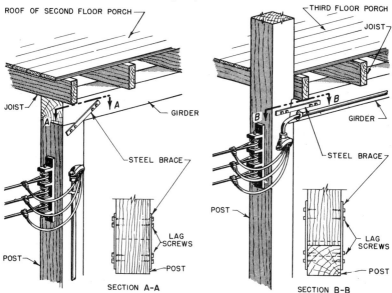

Fig. 4. Pull-off from a
two-story dwelling

Fig. 5. Pull-off from a
three-story dwelling

Many service drops are long and heavy; therefore, the pull-off structure must be sturdy enough to withstand strain imposed under existing weather conditions. If there is any doubt on this point, reinforcement should be provided in a manner best suited to meet these requirements. Several illustrations are given showing how this can be done in situations commonly encountered.

Reinforcing Building. Figure 4 shows the service drop pull-off from the rear porch of a two-story dwelling or apartment building. Two steel braces are bolted or lag-screwed to the post and to the wood girder, as shown.

Figure 5 shows a similar arrangement for a three-story building, the pull-off being made at the third-floor level.

Figure 6 shows a pull-off from a wall section above the opening for a double window in a frame building requiring a heavy service

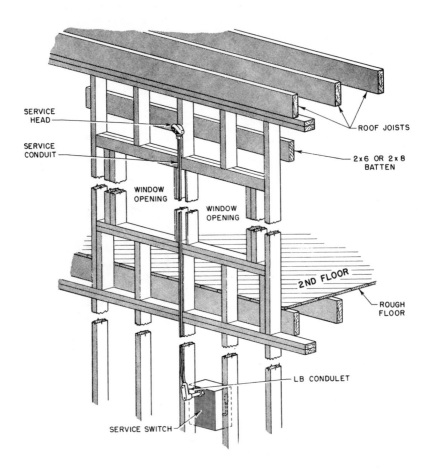

SERVICE HEAD

SERVICE CONDUIT

WINDOW OPENING

WINDOW OPENING

ROOF JOISTS

2x6 OR 2x8 BATTEN

2ND FLOOR

ROUGH FLOOR

LB CONDULET

SERVICE SWITCH

Fig. 6. Pull-off from a wall section

drop. In such cases, a horizontal member is secured to the inner edge of adjacent studdings to transfer some of the strain to them. The conduit must be run down the wall between the two windows.

If the building is not high enough to provide proper clearance for the service drop, a wooden riser 4″x4″ or larger, Fig. 7, is employed to give the necessary height. The riser is located as near as possible to the corner of the building or to a transverse masonry wall in order to get additional support. It must be kept in mind, however, that a building with light walls cannot support a heavy pull. It is safer and better to erect a pole alongside the building in such a case.

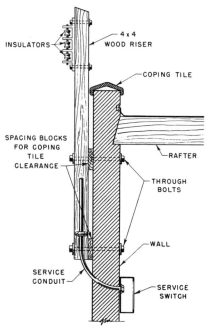

Fig. 7. Pull-off from riser
above the roof

Service Switch

The service switch, Fig. 8, is in most installations a fusible type, the capacity and design of which is suited to the particular application. Thus, it provides both overcurrent protection and disconnecting means. In some cases, the service switch incorporates a meter-fitting and a branch-circuit cabinet.

The NEC lists certain requirements for a service switch: (1) it must be fully enclosed; (2) the operating lever or handle must be on the outside of the enclosure in plain view and within reach of a person standing on the floor; (3) there must be some means for disconnecting supply conductors from their source of electric energy;

(4) there must be a plain indication of whether the switch is in the closed or open position; (5) provision for overcurrent protection is necessary; (6) there must be provision for grounding the neutral conductor; (7) in damp or wet locations, the NEC demands a ¼" air space between the switch enclosure and the mounting surface.

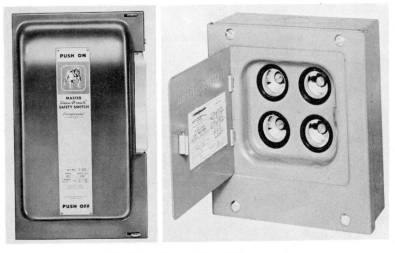

Fig. 8. Service switch and fuse panel
Courtesy of (left) Bulldog Electric Products Co., (right) General Electric Co.

GROUNDING

System Ground

Grounds fall into two separate divisions, namely, the *system ground* and the *equipment ground,* each having a specific duty to perform.

The system ground prevents a rise in voltage beyond that for which the installation is designed. Such increased voltage might occur as the result of lightning, contact of a high-voltage line with wires serving the building, or leakage between windings of the supply transformer. The system ground protects the occupants of a building from high-voltage shock when touching live parts of the wiring. This ground consists of a low-resistance wire or other conducting medium, called the *grounding conductor,* joining the neutral of the incoming service wires to the *grounding electrode.*

Grounding Electrode. The best grounding electrode is the iron pipe coming into the building from an underground water-supply system, not because of water contained in it, but by reason of the large surface area in contact with the earth. Similarly, a metal deep-well casing is good for this purpose.

The Code also approves the use, as a grounding electrode, of an underground gas-piping system because of its large contact area with permanently moist earth. A possible source of danger here is that a poorly installed ground connection may arc if the current flowing through it is strong enough. The arc, which is an electric

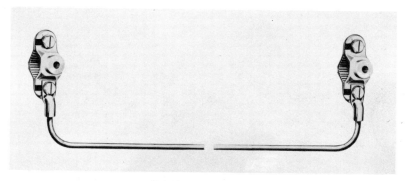

Fig. 9. Meter shunt

discharge across a gap in a circuit, is extremely dangerous when coupled with a leaky gas pipe. Of course, the connection to the ground electrode (no matter what kind it may be) should be made with such fittings, and in such a manner, that *arcing* will not occur under any circumstances.

Connection of the grounding conductor to the underground piping system should be done as near to the latter's point of entry as possible, on the supply side of whatever meter or valve may be located at this point. Where this cannot be done, a bonding jumper, or meter shunt, Fig. 9, should be installed, so the electrical connection will not be interrupted if either meter or valve is taken out. This shunt is placed across the water meter in Fig. 1.

Grounding Conductor. The grounding conductor is attached to the pipe with fittings suitable for the type of conductor, armored cable, conduit, or wire. Figure 10 (right) shows a ground clamp for either bare or armored ground wire. Figure 10 (left) shows a clamp used with conduit. The Code permits bare or insulated wire, either copper or other corrosion-resistant material, solid or stranded, flat or round bus bar, for the system grounding conductor. Its carrying capacity must be proportional to the size of the installation, as indicated in the Code schedule. Clean the grounding electrode thoroughly at the point where the ground clamp is to set, and fasten the clamp tightly.

At the service switch the grounding conductor is connected to the same terminal as the incoming *identified conductor.* The latter must also be fastened to the switch enclosure by means of a lug, a grounding wedge or a grounding bushing. Soldered connections should not be used.

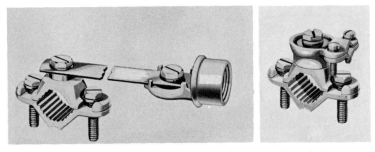

Fig. 10. Ground clamps for use with wire and conduit

Service switches, as well as service-branch-circuit combinations, may have the neutral block mounted directly on the enclosure, instead of being insulated therefrom. The identified service conductor is bolted directly to the neutral block, so that it is not disconnected from the source by opening of the switch. Since it is at ground potential, the bolted connection meets the Code requirement that a means for disconnecting be provided.

Ground Rods. When no satisfactory grounding electrode is readily available, the common practice is to drive one or more rods at such locations and to such depth as to provide a connection having a resistance not higher than 5 ohms.

Where shale, hardpan, or a general rock formation makes it impossible to drive rods, an excavation to permanently moist soil should be made and the electrode buried there. When the excavation for foundation footings is used for this purpose, the electrode should be covered with several inches of earth rather than embedding it in the concrete. By the latter method only a high-resistance ground will result.

A ground rod having a star-shaped cross section, together with its clamp, is shown in Fig. 11. This rod is not only stiffer than a round one of given cross-sectional area, but also has a greater contact surface. This additional contact surface is desirable in localities where dryness or other soil condition tends to produce high resistivity, especially where the capacity of the installation is large or where motors are to be operated at more than 250 volts.

Where soil conditions favor the driving of grounding electrodes, it is permissible to use galvanized-rigid conduit of galvanized-steel

Fig. 11. Grounding electrode
Courtesy of Anaconda Wire & Cable Co.

pipe for this purpose. However, the lower end should be equipped with a driving point and the upper end with a cap, Fig. 12, to protect against crushing under the impact of a sledgehammer during installation. Galvanized conduit which has been given an outside layer of paraffin, lacquer, or other protective substance should not be used, because of the increase in contact resistance resulting from this coating.

Near the seacoast or in other localities where ground moisture is likely to be salty or acid, only rods made of copper or other corrosion-resistant material should be used. The wearing quality of galvanized steel is too uncertain to permit its use under these conditions.

Do not drive the ground rod all the way into or below the surface; let a few inches protrude above the soil so the cap is visible and accessible for inspection of the grounding conductor.

Equipment Ground

The equipment ground eliminates electrical fire hazards and removes the danger of serious injury from electric shock. Such danger arises when a person contacts the metal frame of electrical apparatus that has become *live* due to insulation failure of current-carrying parts.

This ground connection should be applied to the metal frame or enclosure of all electrical apparatus and to conduit or other metallic raceways, regardless of whether a system ground is con-

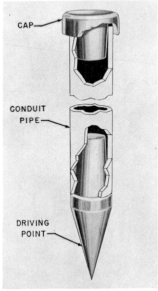

CAP

CONDUIT
PIPE

DRIVING
POINT

Fig. 12. How conduit may be used as a grounding electrode

nected to the wiring at the service switch or not. Wiring systems that are completely metal-clad from service equipment to operating units provide such grounding automatically by way of the metal enclosure.

The NEC requires that secondary windings of transformers supplying interior wiring systems shall be grounded so as to insure that the maximum voltage to ground shall not exceed 150 volts. It recommends that this practice be followed even though the voltage to ground may go as high as 300 volts. This grounding connection is made by the power company. Hence, when electrical contact occurs between a current-carrying member and an ungrounded metal frame or enclosure the latter becomes live.

Inadequate Ground. Consider, for example, an electrical installation in the unfinished basement of a dwelling, where structural materials of the building normally insulate the conduit from contact with the ground. Figure 13 shows a conduit running along the side of a joist, near its lower edge, and down to an outlet on the wall. A gas pipe runs across the bottom of the joists at right angles to the conduit, touching it lightly. If a wire becomes grounded to the raceway at some point in the building, it creates a live conduit. At

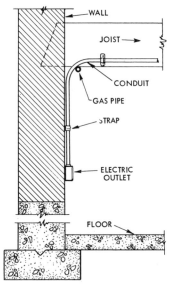

Fig. 13. How current can flow through the ungrounded conduit to gas pipe or other grounded object

the point of contact, current flows from the conduit to the gas pipe, thence to ground, and to the source of supply.

How Fire Can Occur. The flow of current, even though not sufficient to blow the circuit fuse, may be strong enough to set up an arc between the pipes where they touch. In time, the arc may burn a hole in the pipe, resulting in fire or explosion. If pressure between conduit and gas pipe is such as to maintain good electrical contact so that no arc develops, hazard of another sort may still exist. Farther along the gas pipe, say at the meter, electrical resistance may be high because of corroded threads. Resulting temperature rise at this point introduces a fire hazard.

Grounding Conduit System

Grounding of conductors to the conduit system may arise through breakdown of insulation. As pointed out earlier, this creates a live conduit and is a fire hazard. Proper grounding of the conduit system averts this hazard, causing the circuit fuse to blow and thus eliminating the possibility of an appreciable difference of potential occurring between conduit and ground. It should be noted that the NEC permits use of a single conductor for grounding both the wiring system and the service equipment.

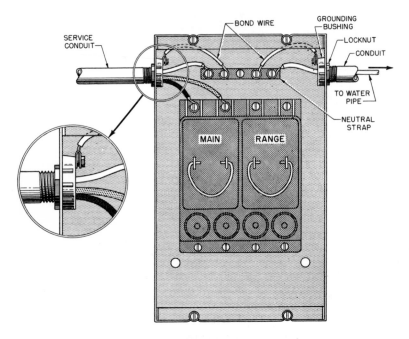

Fig. 14. Bonding conduit at service switch

Bushing. In order to ensure satisfactory electrical and mechanical connection throughout the entire conduit system, it is essential that every length be screwed all the way into its coupling or the threaded hub of its fitting. Where compression-type, setscrew, or indented types of thin-wall conduit couplings or connectors are used, it is necessary that each be thoroughly tightened. Where boxes of the knockout type are entered, the bushing must be screwed onto the conduit as far as it will go, and the locknut drawn up.

Round outlet boxes cannot be used when conduit enters the sides of boxes, since only flat sides provide good seats for bushings and locknuts. Galvanized boxes offer a better ground than do those having an enamel finish, because the enamel, unless completely removed, greatly reduces the metallic contact surface between the box and the faces of bushings and locknuts. The NEC requires that two locknuts shall be employed in the case of rigid conduit, one inside and one outside the box, if the voltage to ground exceeds 250 volts.

Bonding Conduit. The NEC requires that conduits shall be bonded together at the service switch, Fig. 14, instead of depending entirely upon the bushing and locknut connection for the grounding

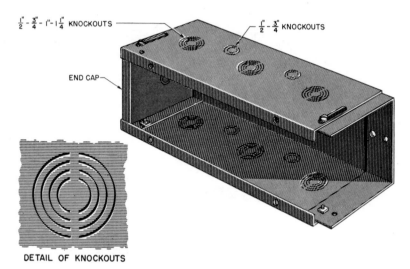

$\frac{1}{2}" - \frac{3}{4}" - 1" - 1\frac{1}{4}"$ KNOCKOUTS

$\frac{1}{2}" - \frac{3}{4}"$ KNOCKOUTS

END CAP

DETAIL OF KNOCKOUTS

Fig. 15. Illustration of concentric knockouts

contact. Bonding may be accomplished by the use of a copper wire or copper strip. *Grounding bushings,* Fig. 14, are also used for the purpose. This bushing has a large screw for securing the ground wire, and a setscrew (not shown) for locking the bushing to the conduit threads. Therefore, it will not become loose as a result of vibration.

Enclosures for service switches and other devices often have what are known as *combination* or *concentric knockouts* to accommodate any of several sizes of conduit without the necessity of reaming out; that is, enlarging a hole or cutting a new one. The arrangement consists of placing the several knockouts one within the other, as shown, in the section of wiring trough, Fig. 15. With these, the wireman removes only those knockouts which provide the size of hole needed for conduit to be used. In this case, unless the entering conduit requires the removal of the largest knockout, a bonding connection must be installed from the conduit to the enclosure in order to restore integrity of the equipment ground, because these knockout rings, even though left in place, may impair the electrical connection to ground.

Grounding Bushing. To do this, install a grounding bushing on each conduit which enters the box through a concentric knock-

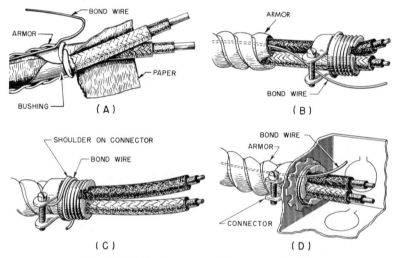

Fig. 16. Methods of grounding armored cable

out, and connect each of the bushings to the enclosure with a piece of copper wire, either bare or insulated. The size of the wire to be used, as given in the code schedule, depends in each case on the overcurrent or fusing protection of the wires entering through the particular concentric knockout. On a large conduit, use a lug for this purpose, installed as previously described. Remove paint or enamel from the box at the point where the grounding connection is to be made.

Grounding Armored Cable

On armored cablework, equipment grounding is done in the same manner as for conduit, except that box connectors are used to secure the cable to the enclosure instead of bushings and locknuts. The code requires that armored cable, except types ACL and ACV, shall have an internal bonding strip, either copper or aluminum, to lower continuity resistance of the outer covering. This wire or strip, Fig. 16A, runs the entire length of the cable just inside the armor; it must be connected to the enclosure or other receptacle into which the cable is entered. There are three ways of making connections:

(1) By drawing the bond wire out through the clamping ears of the box connector, and twisting it around the threaded end of the connector, Figs. 16B and 16C. In this position it is held fast between the enclosure and the shoulder of the connector as the locknut is set up. (2) By running the bond wire into the enclosure along with the wires, Fig. 16D, wrapping it around the threaded end of the connector and clamping it between the enclosure and the locknut. This method requires a washer back of the locknut to make sure that the wire is not damaged by the locknut. (3) In some jurisdictions it is permissible to fold the wire back over the outside of the armor, letting it rest under the clamp, then setting up the latter and cutting off the surplus length.

Any of these methods can be used with setscrew types of box connectors. With methods (1) and (3) the anti-short bushing is slipped into place under the bond wire, Fig. 16A, but in method (2) the anti-short bushing is slipped over the bond wire.

Grounding Nonmetallic Sheathed Cable

On nonmetallic sheathed cablework, the grounding of the neutral housing of a starter, a motor frame, or other current-carrying device, may be done through individual grounding conductors or by means of a grounding conductor which is part of the cable assembly.

Every grounding conductor should be fastened by means of a lug to the metal enclosure or housing in the manner already explained. It is perhaps safer to have a continuous grounding conductor extending all the way back to the grounding electrode at the service location, than to ground each metal enclosure separately to the nearest water pipe.

Grounding is too important to be skimped or slighted.

WATT-HOUR METER

What It Is Used For

When a customer has a 1000-watt appliance and connects it to the electric-light circuit for one hour, the amount of energy used is 1000 watt-hours. A kilowatt-hour is 1000 watt-hours. The watt-hour unit is too small for convenient use, so the kilowatt-hour is used in practice.

The meter that registers the amount of electric energy used by the customer is known as a *watt-hour meter,* even though it records *kilowatt hours.* Since it is usually the only such device in the residence, it is referred to by wiremen and the customer simply as the *meter.* It consists of a small motor the speed of which is directly proportional to the amount of current flowing at any particular instant. The rotating shaft is connected, through a gear mechanism, to dials which record the total number of kilowatt-hours consumed.

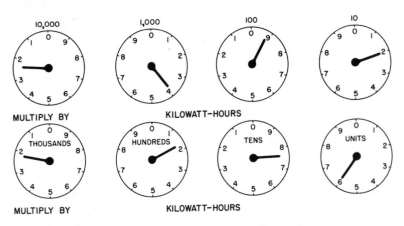

Fig. 17. How to read dial markings on a kilowatt-hour meter

Dial Markings

Each dial is marked off, usually, in ten equal spaces. The major divisions are numbered from *1* to *10,* the figure *0* at the top being used to designate *10.* Two meter faces are shown in Fig. 17. It will be noticed that the numbering of any dial proceeds in the opposite direction from that of its neighbor. This is because pointers of adjacent dials rotate in opposite directions.

Dials are marked by the words *units, tens,* and so on, inside of the ring numerals, as shown in Fig. 17A, or by number *10, 100, 1000,* and *10,0000,* just above the circle as shown in Fig. 17B, to indicate the number of kilowatt-hours each particular one represents. Starting at the right and proceeding to the left, the dials read *units, tens, hundreds,* and *thousands.* Note that each succeeding dial has ten times the value of its right-hand neighbor. On the right-hand dial, each of the numbered divisions has a value of one kilowatt-hour, each complete revolution of the pointer representing ten kilowatt-hours.

Reading a Meter. When reading a meter, jot down the smaller of the two numbers between which the right-hand (units) pointer lies. To the left of the first number, mark down the reading of the 10-dial, and continue with the remaining ones. Thus the readings in Fig. 17 are, respectively, *six, seven, one, two,* the figure being written as: . . . . . *6,* then . . . . *76,* then . . *176,* with the final reading *2176.* Suppose this value represents last month's

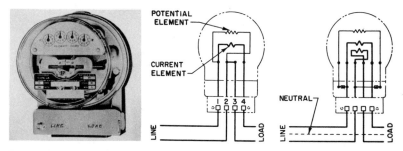

Fig. 18. Bottom-connected meter and how it is used with a 2- or 3-wire supply
Courtesy of Sangamo Electric Co.

reading. Proceeding in the same way, the reading obtained from the upper set of dials is *2392.* Subtracting the former from the latter, *2392 — 2176 = 216,* the consumption during this period was *216* kilowatt-hours.

Care must be taken when a pointer seems to be directly opposite one of the figures, such as the *4* on the upper 1000-dial. The number must be recorded as *3,* because the pointer on the 100-dial at the right has reached only *9,* not having completed the full revolution to move the 1000-pointer to *4.*

With certain meters, the dial reading must be multiplied by a number called the *meter constant,* in order to obtain actual kilowatt-hours. In such devices, there is a notation such as: "Multiply by 10" on the dial or in some other conspicuous place on the meter. Thus, if the dials shown in Fig. 17 were marked *multiply by 10,* the consumption would be *216 × 10,* or *2160* kilowatt-hours.

Meter Wiring

The number of wires brought out for the meter and their arrangement depend on the character of the service, the type of meter, and the kind of load, as well as on standard metering practice set up by the power company. Full information on these points should be obtained from the company before the work is done. Typical residential meter diagrams applicable to various types of underground or overhead electric services are shown in Fig. 18. The unit illustrated here is a *bottom-connected* watt-hour meter having four terminals. The two left-hand ones are connected to the incoming line; the two right-hand ones to the customer's load. The metal cover plate has the words *Line* and *Load* stamped on it.

 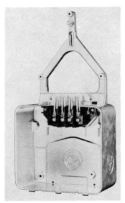

Fig. 19. Meter fitting for bottom-connected meter
Courtesy of Sangamo Electric Co.

Internal Meter Diagrams. Current elements in the meter diagrams are represented by the heavy lines, potential elements by the lighter lines.

In the 2-wire meter, a supply wire enters at terminal No. 1 and connects to both potential and series windings. The other end of the series coil goes to terminal No. 4, and thence to the load. The remaining supply, where it connects to the potential coil before continuing, enters terminal No. 2, goes through a short connection to terminal No. 3, and out to the load.

In a 3-wire meter the two outside wires pass into and out of the meter, the potential winding bridging them, as shown. The neutral wire goes directly to the load.

Meter Installation

Bottom Connected Meter. The simplest kind of meter installation is a bottom-connected unit, Fig. 19, mounted just above the service switch. An opening is provided in the top of the switch to fit the bottom of the meter. After connections have been made, a metal

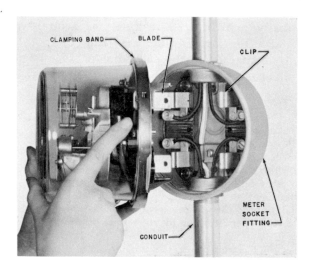

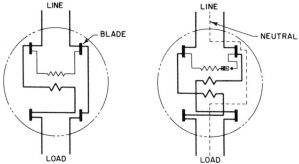

Fig. 20. Socket-type meter and diagrams for use with 2- or 3-wire supply

Courtesy of Sangamo Electric Co.

cover is placed over the meter terminals and sealed by the power company.

Socket-type Meter. Power companies required meters that could be quickly installed or removed. The socket-type meter, Fig. 20, was developed to meet this need. The internal construction and diagram are the same as for other types. The terminals are in the form of blades which fit into clips, as shown. The meter-socket fitting is usually supplied by the power company. The wireman must fasten it to the building, install the service conduit (or service-entrance cable) and wires, and then attach them to the clips in the fitting.

Supply wires are connected to the upper terminals, and load wires to the bottom. In a 3-wire meter the neutral wire usually passes through the fitting without making contact. Sometimes, however, it is attached to the enclosure in order to ground the fitting without use of a separate grounding conductor.

REVIEW QUESTIONS

1. What are service drop wires?
2. Who decides where the service drop shall be located?
3. How much air space is required between a switch and a damp wall upon which it is mounted?
4. To what height on a riser pole does NEC require service wires to be protected?
5. Who connects the service wires in a sidewalk handhole?
6. How is proper clearance for the service drop obtained in buildings of low height?
7. Does the electrical contractor usually run the underground service wires direct to the manhole in the street?
8. Name the two divisions or kinds of grounding.
9. What is the most desirable electrode for a system ground?
10. What is the purpose of a bonding jumper or meter shunt?
11. Where should the system grounding conductor be attached?
12. In system grounding, what is the maximum permissible resistance when ground rods are used?
13. Why should a few inches of driven ground rod protrude above the soil?
14. What is the purpose of an equipment ground on electrical apparatus?
15. When are double locknuts required by NEC?
16. In what two ways may a poor ground connection be the cause of a fire?
17. Does the enamel finish on conduit provide as good a ground as galvanized material?
18. Why should a bonding jumper be required with concentric knock-outs?
19. Why is a bare metal ribbon or wire provided inside the steel covering of armored cable?
20. What is the simplest kind of meter installation?

Chapter 5

Lighting Outlets and Switches
Bell Wiring
Intercom Systems

QUESTIONS THIS CHAPTER WILL ANSWER

1. What are 3-way and 4-way switches?
2. How are they wired?
3. What is an electrolier switch?
4. How are chimes, door openers, and number lights connected?
5. How are intercom systems wired?

SINGLE-POLE AND DOUBLE-POLE SWITCHES

Elementary Procedures

This section is concerned mostly with the control of lighting outlets. However, there is no reason why any other type of load which requires merely the closing or opening of its supply wire may not be controlled in the same manner as the lamp shown in the illustrations.

The figures in this chapter show *polarity wiring*, where one of the wires entering an outlet box is pictured as a dotted line, all others as heavy lines. The dotted line indicates the neutral or identified conductor which is grounded; it is always white or natural gray for residential circuits. The neutral wire is connected to the screw shell of the lamp socket, thus making it impossible for anyone touching the shell or the lamp base to receive a shock.

An important rule is that the neutral wire should never be run to a switch; only one of the *outside,* or *hot,* wires should be used for this purpose. The hot wires, indicated by heavy lines in the illustrations, may be red or black. Other colors, except white or green, are sometimes allowed. Where two or more wire ends are brought together within an outlet box, a pigtail splice is made.

Many faulty connections can be avoided if a simple method is followed when fastening wires to switches, receptacles, and sockets. Since wiring to electrical equipment is always attached with right-hand screws, the wire should be bent around the screw in a clockwise direction so that tightening of the screw will draw the wire in

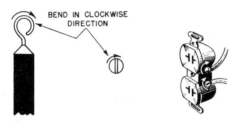

Fig. 1. Method of placing wires under screws prior to tightening

and not push it away. Fig. 1 shows the correct method of placing wires under screws prior to tightening.

Reasons for Polarity Wiring. Neutral conductors should be run in the same conduit or cable as "hot" conductors of the circuits to which they are connected. There are two reasons for this practice: (1) to prevent overloading of neutral wires through wrong connections, and (2) to avoid harmful effects of induction. The first of these principles may be understood with the help of Fig. 2. Conductors for the two two-wire circuits are No. 14 Type R copper. The hot wires of circuits *1* and *2* each carry a load of 15 amps, which represents their maximum allowable current-carrying capacity.

To Prevent Overloading of Neutral Wires. If the lamps are properly connected, Fig. 2A, each neutral wire carries 15 amps. But, if the lamps of circuit *2* are connected accidentally to the neutral conductor of circuit *1,* as indicated in Fig. 2B, this wire is forced to handle 30 amps. Such an overload damages insulation on this particular wire, and the high temperature of the overheated copper may damage insulating coverings on adjacent conductors.

When the neutral conductor is run with the hot conductor of the circuit, this sort of accident is not so likely to happen. Overload may still occur, as a result of carelessness, where two or more circuits are placed in the same conduit. Nevertheless, such a dangerous situation is not so likely to arise as when neutral con-

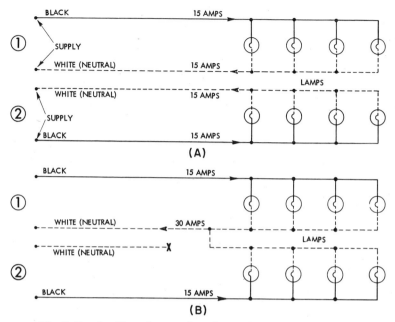

Fig. 2. Overloading of neutral conductor by wrong connection

ductors are installed without regard to pairing them with the hot ones.

To Avoid Harmful Effects of Induction. In order to grasp the second principle, the term *induction* must be understood. Figure 3A shows two conductors which make a loop around a piece of iron, an A-C current flowing in them. One result of the iron core is to create, through magnetic action, a voltage which opposes the supply, thus increasing voltage drop in the circuit. The second result of magnetic induction is severe heating of the iron itself.

Figure 3B illustrates a pair of circuit conductors inside an iron conduit. Note that the iron is outside the wires instead of being enclosed by them. Also, the magnetic flux set up by current flowing in one conductor is offset by current which flows in the opposite direction through the second conductor. Magnetic effects are thus cancelled out, and induction cannot take place.

Where the hot conductor is in one conduit and the neutral conductor in another, between junction boxes *J-1* and *J-2*, Fig. 3C, the wires enclose an iron core which is formed by conduits and

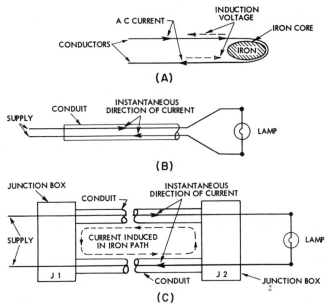

Fig. 3. Effect of induction

junction boxes. As a result, induction takes place. Voltage drop in the circuit is increased, and the iron members are heated. The greater the amount of current in the circuit, the greater the voltage drop and the heating. With currents of a few amperes, heating effects are not serious. But with larger currents, say upward from 30 amps, dangerously high temperatures may be reached.

Single-Pole Switches

A simple, single-pole *switch-loop* is shown in Fig. 4. The outlet to be controlled is at the end of the supply circuit. There are three 2-wire pigtail splices in the outlet box, one connecting the neutral supply wire with the fixture wire coming from the screw shell of the socket, one connecting the hot supply wire to the *switch leg*, and one connecting the *return leg* from the switch with the fixture wire coming from the center contact of the socket. On a conduit job, one splice may be saved by pulling through enough wire of the hot leg to reach the switch.

A similar switch leg is shown in Fig. 5, except the supply circuit to the outlet does not stop there but goes on to additional outlets which are independent of the switch. The two wires from

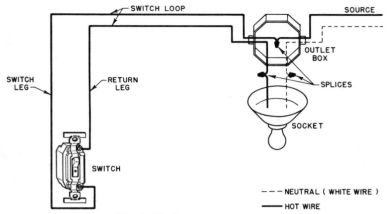

Fig. 4. Single-pole switch loop

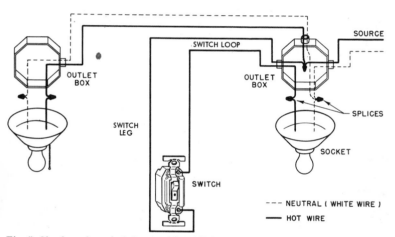

Fig. 5. Single-pole switch loop with additional outlets independent of switch

the outlet box to the switch are called the *switch loop*. The wire feeding the switch is the *switch leg*. There are two 3-wire pigtail splices and one 2-wire splice in the outlet box. The reason for two of the splices being 3-wire is that two wires of the circuit must continue on to supply additional outlets. The incoming neutral, the outgoing neutral, and the fixture wire coming from the socket screw shell are joined. The incoming hot wire, the outgoing hot wire, and the switch leg are connected together. The return leg from the switch and the fixture wire coming from the socket center contact are joined.

Two-Gang Switch. A two-gang switch, that is, two switches in one box and under one faceplate, each controlling a separate outlet, is shown in Fig. 6. In this case the supply circuit enters outlet box No. 2 and from there goes into outlet box No. 1. In box No. 2 there is a 3-wire splice for the neutrals, a 2-wire splice for the incoming and outgoing hot leg of the circuit, and a 2-wire splice for the return leg from the switch. The return leg must pass through

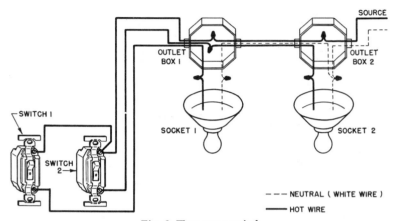

Fig. 6. Two-gang switch

outlet box No. 1 to reach the switch; a 2-wire splice is shown in this box. This splice can be saved by pulling through enough wire to reach either the switch or outlet box No. 2, depending on the direction of pull. On a knob-and-tube job, the return leg from switch 2 does not enter outlet box No. 1 at all but runs directly to outlet box No. 2. Outlet box No. 1 has, in addition to a splice for the return leg from switch 2 to outlet box No. 2, two 2-wire splices for the wires of the switch loop. There is also a 2-wire splice for the neutral and load wires.

A somewhat different arrangement is shown in Fig. 7, where the switch loop is taken from outlet box No. 2, instead of No. 1, as in the preceding case. There is one additional 2-wire splice in outlet box No. 2 but only two 2-wire splices in box No. 1, a saving of one 2-wire splice. Again, if this were a knob-and-tube job, the return leg from switch 1 would run directly to box No. 1 instead of going through box No. 2, as shown.

Three-Gang Switch. A 3-gang switch for the control of three separate outlets fed by a circuit should be arranged in the same manner, except that there would be three return legs from the switches. The third return leg runs to the third outlet box through each of the other two boxes.

Double-Pole Switches

For many switching jobs, the single-pole unit is adequate because it is connected so as to interrupt current flow in the hot wire and

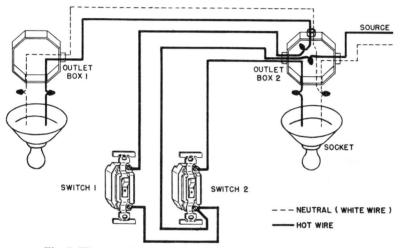

Fig. 7. Two-gang switch with switches placed between outlets

leave the fixture *electrically dead* when it is in the "off" position. Sometimes it may be desirable to break both sides of the line, in which case a double-pole switch must be used. As an example, Fig. 8 shows a 3-wire convenience outlet and a double-pole switch that controls both hot line wires, the neutral being carried through unbroken to the outlet.

Many plug-in types of appliances use double-pole switches as a safety measure because the manner of inserting the plug determines which shall be the hot wire inside the device.

Three-Way Switches

Description. There are times when it is desirable to control one or more lamps from two different points; for example, in a long corridor, a stairway, or a room with two separate entrances. This is done by means of two 3-way switches. Instead of having a single terminal at both top and bottom, there is an additional terminal

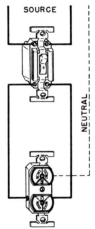

Fig. 8. Double-pole switch used to control both
line wires supplying a convenience outlet

at one end. In order to identify these terminals throughout this discussion, the single terminal at one end of the base is called the *hinge point*. Some manufacturers use a black oxide screw here for purposes of identification. The hot leg is connected to it. The

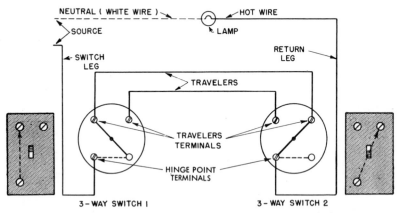

Fig. 9. Illustrating the operation of a three-way switch

terminals at the other end of the base, Fig. 9, are connected to similar terminals of the other control switch by two wires which are called *travelers*.

Construction of Three-Way Switches. Two blades of the switch mechanism are so arranged that the hot terminal is connected to either one of the traveler contacts as the handle is moved from one position to the other. That is, the switch itself has no "off" position. The hinge point of one switch is connected to the hot side of the supply circuit; the hinge point of the other switch is connected to the center contact of the lamp to be controlled. The screw shell of the lamp is connected to the neutral supply wire.

In the illustration, which is a diagram of the connections for a single outlet controlled by two 3-way switches, a round-base, rotary switch is employed to simplify the discussion. The standard rectangular-base switch with the terminals in the corners of the base is shown to the left and right of it, Fig. 9.

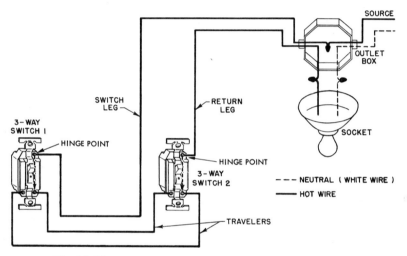

Fig. 10. Three-way switch circuit controlling a single outlet

In the rotary switch a movable bar connects a base terminal to one of the travelers. The contacts at the bottom are permanently connected, as indicated by the dotted lines, thus forming the hinge point. No matter in which position the switch is turned, one or the other of the traveler terminals is made live. In the tumbler unit the electrical connections are changed in a similar manner by moving the handle of the switch.

Operation of Three-Way Switches. The path for the current, Fig. 10, may be traced from the *hot* supply wire in the outlet box to the hinge point of switch *1*, then to the right-hand traveler terminal, and through the upper traveler to the left-hand traveler terminal of switch *2*, where the circuit is broken. Because both switch levers are in the same position and there is no connection from the left-hand traveler terminal to the hinge post, the lamp does not burn. But, if the lever of switch 2 is moved to the other position, current will flow from its left-hand traveler to the hinge point, through the return leg to the center contact of the socket, and through the lamp to the neutral. The lamp will now burn.

If the switch levers are returned again to the position shown in Fig. 10, the lamp goes out. Suppose the lever of switch *1* is moved to the other position. From the hinge point the flow is to the left-hand terminal, through the bottom traveler to the right-hand

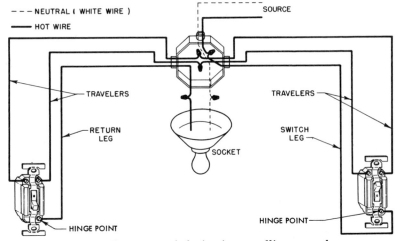

Fig. 11. Three-way switch circuit controlling an outlet
from a point on either side of the outlet

terminal of switch 2 and to the hinge point, through the return leg to the center contact of the socket, and through the lamp to the neutral, causing the lamp to burn. Thus, the movement of either switch lever controls the lamp.

The arrangement of wires, Fig. 10, must be followed if the wiring is done in conduit or armored conductors. However, if either non-metallic sheathed cable or knob-and-tube is used, the switch leg can be run direct from switch *1* to the outlet instead of by way of switch 2.

The scheme for controlling a lamp from two points at opposite directions from the outlet is shown in Fig. 11. Three wires are run to each of the two switches, the travelers being connected in the outlet box.

Controlling Several Outlets. A method for controlling several outlets from two points is shown in Fig. 12, the supply circuit entering outlet box 3. One of the switch loops is taken out of box No. 1, the other out of box No. 4 because these two are close to the switches.

The circuit path may be traced from the point of entry in box No. 3, through all the outlets, and back to the starting point. The neutral goes to the screw shell of each of the sockets, and the *hot* leg through boxes No. 3 and No. 4 to the hinge point of switch 2. Assume that in both switches, at the moment, hinge points are making contact with right-hand traveler terminals. The path continues from the hinge point of switch 2 to the right-hand traveler which is the uppermost conductor, Fig. 12, through the four outlet boxes. Descending toward switch 1, this wire connects to the left-hand traveler terminal. The path is broken here, making the circuit incomplete, because the hinge point of the switch is making contact with the other traveler.

If switch 1 is now manipulated, current may flow from switch 2 over the upper traveler to the hinge point of switch 1, through the return-switch leg to the center contacts of the four sockets, and through the lamps to the neutral leg of the supply circuit. Or, leaving switch 1 in its original position (lamps not burning), operate switch 2 to connect its hinge point with the left-hand traveler. Current then flows over this wire through the four outlet boxes to switch 1 and to the hinge point, over the return switch leg to the lamps, and through the latter to the neutral supply wire, causing them to burn.

In a metal-clad system, all wires must be run in the same conduit or armor from outlet to outlet. On knob-and-tube or nonmetallic cable installations, wires can be run in any desired manner. For example, it may be convenient to carry the travelers direct from switch to switch instead of through the four outlet boxes.

Where conduit is employed, two splices can be eliminated in each of the outlet boxes, Nos. 1, 2, and 3, by travelers in these boxes. This is done by pulling the travelers through the boxes without cutting them. In outlet box No. 4, the travelers and the switch leg can be drawn through in this way, thus saving three splices. This should be done whenever possible, that is, without paying too high a price in labor. For example, the outlets may be far apart or isolated from each other by various partition walls, so that more

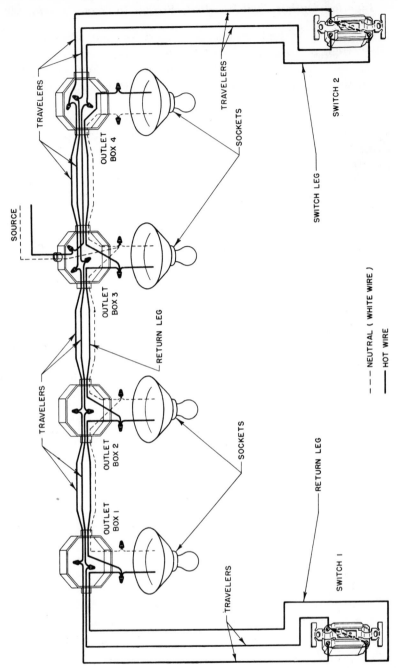

Fig. 12. Three-way switch circuit controlling several outlets

time would be lost returning from one outlet to the other for pulling back the unused slack than would be required for making splices.

Note in Fig. 12 that there are only three hot wires (including the travelers) from box No. 3 to boxes No. 1 and No. 2; whereas, there are four between box No. 3 and box No. 4. This additional leg means not only more wire and larger conduit, but also more labor than for the other two spans. This is the reason for selecting switch *2* as the *hot* line point, rather than switch *1*.

Four-Way Switches

Construction. When it is desirable to control outlets from three points, a 4-way switch is placed in circuit with the two 3-way switches. A 4-way switch is similar in size and appearance to a 3-way

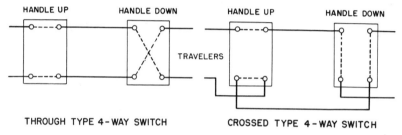

Fig. 13. Two types of four-way switches

of the same general type, except that it has four terminals. Many of the 4-way switches have the terminals arranged in the same relative positions as those of a double-pole switch, Fig. 8. The latter is always recognized by the words ON and OFF that mark positions of the handle.

There are, in general, two classes of 4-way tumbler switches, Fig. 13. One is known as the *through type,* the other, the *crossed type.* In the first type, travelers from a 3-way or 4-way on the left connect to terminals on that side. Travelers going to the other 3-way or 4-way on the right connect to terminals on that side.

When the handle of the switch is actuated, the electrical connections inside the switch change from straight across to diagonally across, or vice versa, as shown in the left illustration of Fig. 13.

In the crossed type, one of the travelers coming in from the left goes to a switch terminal on that side. The other one crosses over to the terminal on the right. Also, on the outgoing travelers one connects on the right side, while the other crosses over to the left. Operation of the switch handle changes inside connections from straight across to up-and-down, or vice versa.

The wireman may encounter a 4-way switch of the through type in which the connections, instead of going directly through from left to right, left view of Fig. 13, go from top to bottom. When installing new switches, check the diagram on the box to make sure what has to be done.

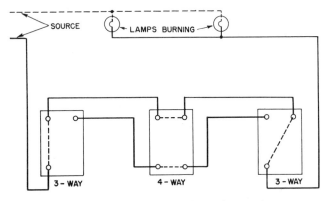

Fig. 14. Controlling outlets from three points

Operation of Controls at Three Points. In order to understand operation of controls from three points, it is well to spend some time analyzing the circuit in Fig. 14 at the moment current flows through wires and switches as indicated by the dotted arrows, causing the lamps to burn. Suppose that the left-hand 3-way switch is operated. Mark this connection on a piece of paper, and trace the circuit through the switches. Note that the lamps go out because there is no connection at the right-hand 3-way switch from the left traveler wire to the lamps.

Operation of Switches at Four Control Points. Next, suppose that the 4-way switch is snapped, making connections in the criss-cross manner shown in the right-hand view of Fig. 13.

Tracing the circuit, current flows to the hinge of the left-hand 3-way switch, diagonally to the upper right contact and the lower traveler, then to the lower left terminal of the 4-way switch. The path continues diagonally to the upper right terminal, along the upper traveler to the right-hand upper terminal of the second 3-way switch, then diagonally to the return leg, and through the lamps, completing the circuit.

If the right-hand 3-way is snapped, current flow ceases. The circuit may be restored, however, by turning either of the remaining switches. A profitable exercise for the learner is to make rough sketches, similar to Fig. 14, of resulting current paths when the three switches are operated in the various possible ways and combinations.

A wiring diagram involving four control points is shown in Fig. 15. Two crossed-type 4-way switches and two 3-way switches are em-

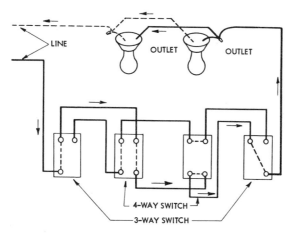

Fig. 15. Wiring arrangement for controlling outlets from four places

ployed. At the instant shown in the diagram, current flows through the first 3-way switch to the upper traveler, to the 4-way switch, where it crosses over the traveler from the opposite direction, then down to the bottom traveler, and along it to cross over at the second 4-way switch. Here, it continues along the bottom traveler to the left contact of the second 3-way switch, then to lamps that are in parallel, and through them to the neutral wire, completing the circuit.

If any one of the four switches is turned, the path is broken. Continuity may be restored by operating any one of the other three. In this instance, too, the learner can profit by making rough sketches of current paths when these four switches are operated in numerous sequences.

To control a circuit or a group of lamps from any greater number of points than shown in the examples, all that is required is to add 4-way switches. Thus, three 4-way switches are required if there are five control points, four if there are six points, and five if there are seven points. There must be a 3-way switch at each end of the control circuit, with the required number of 4-way switches in between.

OTHER TYPES OF SWITCHES

Electrolier Switches

Electrolier switches are used with fixtures that have two or more lamps, or groups of lamps, in dining rooms or similar locations, where different intensities of illumination may be wanted for vari-

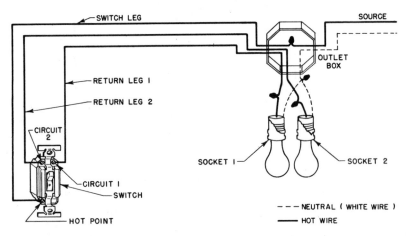

Fig. 16. Two-circuit Electrolier switch

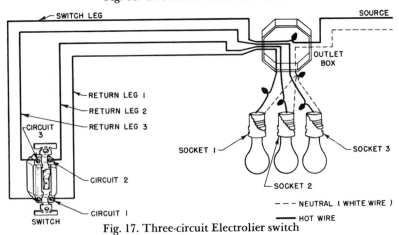

Fig. 17. Three-circuit Electrolier switch

ous occasions. The switch mechanism is actuated by repeated manipulation of the button or handle. Switches offering various combinations may be obtained by specifying the kind wanted. Usually they are either 2-circuit or 3-circuit. In appearance they resemble standard units and can easily be mistaken for 3-way, 4-way, or double-pole switches. Figure 16 shows a wiring diagram for a 2-circuit switch, and Fig. 17 that for a 3-circuit switch. In either case there is a hot-switch leg from the outlet to be controlled to the common terminal of the switch and a return switch leg for each circuit from the switch to the outlet box. Thus, three wires are used for a 2-circuit and four wires for a 3-circuit switch.

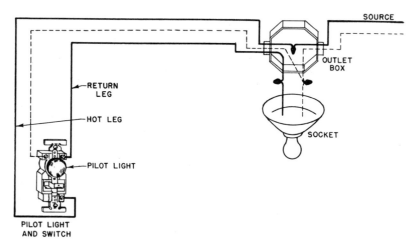

Fig. 18. Illustration of a pilot-light switch

Pilot-Light Switches

These switches are used for controlling lamps in pantries, furnace rooms, or other infrequently entered locations. They are employed on electric irons, hot plates, and such devices which may be damaged if connected accidentally for too long a period. The pilot light burns as long as the switch is closed. Figure 18 shows how one of these switches controls a lighting outlet, but the connections to a heating outlet are exactly the same. To help in making proper connections some pilot light switches have the neutral terminal identified by a white or silver-colored screw, and the *hot-one* by a colored screw.

SPECIAL NOTE: White Wire in Switch Loop

As a general rule, the neutral conductor (white wire) should not be switched or used as one of the legs in a switch loop. This rule is readily applied with respect to either knob-and-tube or conduit work.

However, armored and nonmetallic sheathed cable, whether 2-, 3-, or 4-conductor, have a white conductor. The Code now states that cable having an identified conductor may be used for single-pole, three-way, or four-way switch loops where connections are made so that the *unidentified* wire is the return conductor from switch to outlet. This rule nullifies the old requirement that the identified conductor must be painted at the ends to destroy identification.

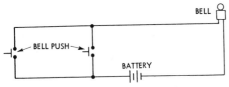

Fig. 19. Simple bell circuit

BELL WIRING

Simple Arrangements. Bell wiring should be thoroughly understood by the inside wireman. Figure 19 shows a simple bell circuit which consists of a battery, a bell, and a set of bell pushes.

The bell push, also known as *push button,* is in series with battery and bell. When contacts of the button are pressed together, current flows from the negative terminal of the battery to the bell, then through the connecting wire to the button, and through the button to the positive terminal of the battery. Any number of push buttons may be arranged in parallel across the connecting wires at the left side of the battery.

In the return-call system, Fig. 20, there is a bell and a push button at point *A,* and a like combination at point *B.* This method of connection is used for signalling between points which are remote from one another. If the button at *A* is pressed, the bell at *B* rings but not the bell at *A.* Likewise, when the button at *B* is pressed, only the bell at *A* rings.

The paths may be traced readily. Starting at *B,* when the button is pressed, current from the negative terminal of the battery flows through the wire to the button, through the closed contacts and the connecting wire to bell *A*, then through the bell and wire to the positive terminal of the battery. The bell at *B* does not ring becaues its circuit is open at the push button contacts of *A*. In like manner, when the button at *A* is pressed, current flows out of the negative terminal of the battery through bell *B* and connecting wire to push button *A*, the contacts of which are closed. The current then returns through the wire to the positive terminal of the battery.

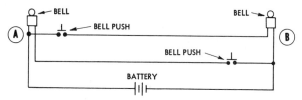

Fig. 20. Return-call bell system

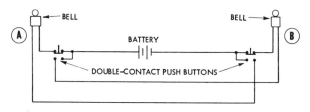

Fig. 21. Return-call bell system using double contact push buttons

The circuit of Fig. 21 makes use of double-contact push buttons in a return-call arrangement. These push buttons are constructed so that the lever is normally held against an upper contact by means of a flat spring. When the button is pressed, the lever opens the upper contact, and makes the lower one. When button *B* is pressed, current flows out of the negative terminal of the battery through the connecting wire and lever to the lower contact. After passing through the wire to bell *A*, it returns through the upper contact of button *A*, the lever which is still pressed against the contact, and the circuit wire, to the positive terminal of the battery. Bell *B* cannot ring because its circuit is broken when push button *B* is pressed.

When push button *A* is actuated, current flows out of the negative terminal of the battery through the contact lever of button *B*, upper contact of bell *B*, connecting wire, and the lower contact of button *A*, to the lever which is being held in contact with it, and from there to the positive terminal of the battery.

Chime Circuits. Old style bells have been largely displaced in the modern home by door chimes. In the more common type, two distinct tones are provided so that the householder may know whether the caller is at the front door or at the back. The front button usually brings forth a succession of tones, or a chord, while the back door button gives rise to a single tone.

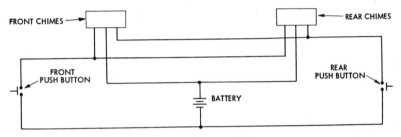

Fig. 22. Wiring for two-tone chimes with battery power supply

The diagram, Fig. 22, shows two sets of chimes, one at the front and one at the back. Although this is a common arrangement, a single chime, centrally located so that it may be heard all through the house, is sometimes employed. The arrangement is the same in either case, the two units being connected in parallel as shown. In fact, three or four chimes may be so connected, when the power supply is large enough to handle them.

The chime unit has three lead wires, the center one of which connects to both of its magnetic coils or solenoids. When the front button is pressed, current flows out of the negative terminal of the battery through the button, the front door solenoid of the chime, and the connecting wire, to the positive terminal of the battery. Since two units are connected in parallel, the front door notes will be sounded by each of them. If the rear button is pressed, current from the negative terminal of the battery will flow through the back button, the rear-door solenoid in both chimes, and the common wire, back to the positive terminal of the battery.

Batteries are seldom used today for these systems. Figure 23 illustrates a circuit which employs a transformer power supply. Connection of the chimes is exactly the same as in Fig. 22, but there are two additional items which are frequently used. The door opener is wired so that it may be operated from a push button at the back of the house as well as near the front. If the circuit is traced, it may be seen that the door opener is in series with the transformer and either of the push buttons in much the same manner as an ordinary doorbell.

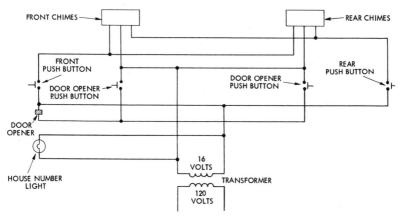

Fig. 23. Wiring for two-tone chimes, door opener and
number light with transformer power supply

The house number light is connected permanently across the terminals of the transformer. There is no switch in this circuit, so that the light burns continuously. The globe consumes very little power, and its output is so feeble that it is unnoticed in daylight hours.

INTERCOMMUNICATING TELEPHONES

Two-Station Arrangement. Intercommunicating systems, called "intercom systems" for short, are used in the home for conversing between widely separated rooms, such as from bedroom to kitchen.

A two-station system is illustrated in Fig. 24. The equipment at either station is identical, consisting of a transmitter, a receiver, a hook switch, a bell, and a push button. The transmitter and receiver at each station are connected in series. When the hooks are raised so the levers make contact with the lead wire from the receiver, the two sets of transmitters and receivers are connected in series with each other and with the battery.

A two-way conversation may be carried on, vibrations of the transmitter diaphragm at either end causing rapid fluctuations in the battery current, and reproducing sound of the voices in both receivers.

When the hooks are down, as indicated in the figure, the bells, push buttons, and battery comprise a return-call system. Note that the receiver must be left on the hook while the button is actuated. If this is not done, the bell at the other end cannot ring, because the battery circuit will be open at the hook switch.

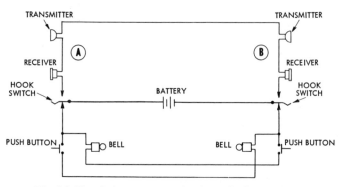

Fig. 24. Simple intercommunicating telephone circuit

Four-Station Intercom System. The selective-ringing, common-talking arrangement of Fig. 25 is widely used in residences and apartment houses. The individual station equipment is similar to that for the simple intercom system of Fig. 22, except that multiple ringing buttons or keys are employed instead of single push buttons. When any two receivers are lifted from their hooks, a two-way telephone conversation may be carried on between the stations.

The battery connection is different from that in the simple intercom circuit. The positive terminal is connected to a common wire that leads to the hook switches. The negative terminal is attached to a retard coil. The opposite terminal of the retard coil is fastened to the common wire that runs to the transmitters. The battery furnishes a steady exciting current to stations with hook switches in the raised position.

The retard coil offers little resistance to the steady battery current, but very high resistance to the passage of high-frequency voice current. If the retard coil were not present, most of the high-frequency current variations originating at one station would pass through the battery, which has a low resistance, rather than through the transmitter-receiver circuit of the other station, which offers a comparatively large resistance to them.

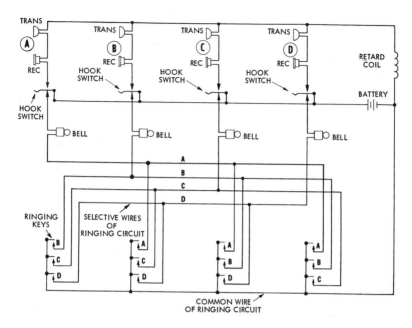

Fig. 25. Four-station intercom system—selective ringing—
common talking

The path of the steady exciting current is from the negative battery terminal of the retard coil, to the upper common wire, then through the transmitters and receivers of the active stations and the raised hook switches to the lower common wire, and from there to the positive terminal of the battery. The path of the high-frequency current variations set up by the party talking at station *A*, for example, is done through the receiver to the hook switch, along the common wire to the other talking station, *C* for example, upward through the receiver and transmitter to the other common wire, and back to station *A*. If station *C* talks to station *A*, the path is reversed.

When station *A* decides to talk with station *C*, key *C* is pressed at station *A*. Current flows from the negative battery terminal to the common ringing wire, to ringing key *C* at station *A*, through selective ringing wire *C* to the bell at *C*, from there to the hook switch at *C*, and through the common wire to the positive battery terminal, completing the circuit so that bell *C* rings. As *C*'s receiver is lifted, the talking circuit is completed at *C* and the ringing circuit is opened at that point.

The ringing circuit consists of a selective ringing wire for each station and a common wire. Connection is made at a station to each of the selective ringing wires except its own. The two wires of the talking circuit should be a twisted pair in order that interference from electrical devices or equipment in the vicinity will be minimized.

With this particular system, any one or both of the other stations may listen in on the conversation, because all are connected in parallel. In order to provide selective talking, it would be necessary to add a dial arrangement together with selective talking wires from each station to every other station.

Fig. 26. Modern combination unit

Courtesy of NuTone, Inc.

Combination Loud-Speaker/Intercom Systems. Combination units such as that in Fig. 26 have recently become popular. The system consists of a master station and a number of remote stations in various rooms. Stations are often provided in vestibules or entries so that the householder may talk to a caller without opening the door. The system illustrated here combines choice of radio or phonograph music with a selective-ringing, selective-talking intercom system. Connection of these devices by the wireman is relatively simple because the manufacturer furnishes color-coded wire, numbered terminal blocks, and detailed instructions. The principles involved are the same as those outlined above.

REVIEW QUESTIONS

1. In what direction should one twist a wire around a terminal screw?
2. What is meant by the term "switch leg"?
3. What is a switch loop?
4. Is the neutral wire called the "return leg?"
5. How many wires pass from an outlet box to a two-gang switch?
6. Can three-way switches be used to control plug receptacles?
7. Is the neutral wire usually connected to a switch?
8. What are "travelers"?
9. How can one tell the difference between a double-pole switch and a four-way switch?
10. Could a double-pole switch be used to control two circuits?
11. Could a four-way switch be used as a three-way switch?
12. Is it customary to use a three-way and a four-way switch to control a light from two points?
13. Must the neutral wire be connected to a two-pole switch that controls separate halves of a plug receptacle?
14. How many wires pass from a three-way switch to a four-way switch?
15. How many three-ways and how many four-ways are required to control a light from four places?
16. Where are pilot-light switches used?
17. Can ordinary push-buttons be used in a return-call bell system?
18. Are door openers supplied from separate transformers?
19. Are residential intercom systems usually selective-ringing, selective-talking?
20. How many selective ringing wires are needed for a six-station intercom system?

Chapter 6

Methods of Wiring

Knob and Tube

QUESTIONS THIS CHAPTER WILL ANSWER

1. How do you install open knob-and-tube wiring?

2. Where is the use of open knob-and-tube wiring advisable?

3. What are the basic NEC rules dealing with concealed knob-and-tube wiring?

4. What is the general procedure for installing concealed knob-and-tube wiring in a new house?

5. How do you install concealed knob-and-tube wiring in an old house?

ADVANTAGES OF KNOB-AND-TUBE WIRING

Knob-and-tube wiring is the oldest type of electrical installation which meets requirements of the National Electrical Code. Today this method is not approved by many local codes.

Knob-and-tube wiring has a definite advantage for installations in damp or wet locations and also in buildings where certain corrosive vapors exist. Temporary installations—for example, fair grounds and construction jobs—can be more readily served by knob-and-tube wiring. It should be noted that operating temperature will be lower than that of other wiring systems because of better ventilation afforded the conductors.

This method of wiring is divided into two types, namely, *open* and *concealed*. It is said to be open or exposed where the wires are run upon surfaces of walls and ceilings. It is concealed when hidden between ceiling and floor, or run on joists in an inaccessible attic.

The exposed type of wiring is suitable for barns, sheds, and factories where appearance is not an important factor but where ease of servicing or altering is important.

OPEN KNOB-AND-TUBE WIRING

Insulators

The rules for open work are simple and easy to follow. Insulators should be free from checks, rough projections, or sharp edges which might damage insulation. All knob-and-tube wiring must be done with single conductors. Wires size No. 14 AWG to No. 10 AWG are supported usually on 2- or 3-wire cleats, Fig. 1, and secured by screws or nails. Leather-protecting washers are used with nails. The nails or screws should extend into the wood at least as far as the thickness of the cleat. Wires size No. 8 AWG and larger are best supported on single-wire cleats, Fig. 2, and secured by two screws. Direction of joists must be considered when planning the wiring installation, because lath and plaster do not provide sufficient support for insu-

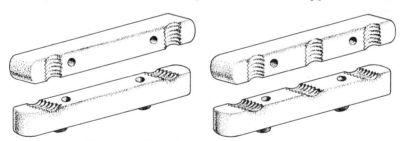

Fig. 1. Standard cleats for two and three wires

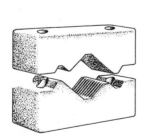

Fig. 2. Standard one wire cleat
Fig. 3. Solid porcelain knob
Courtesy of General Electric Supply Corp.

lators. The National Electric Code demands a support every 4½ ft. with a joist spacing of 12″ or 16″. An insulator is required on every third or fourth joist.

Surface Clearance. Cleat insulators should be of sufficient height to provide ½″ clearance for the surface wired over in dry locations. In damp or wet locations, a clearance of 1″ is required, so that insulating knobs must be used instead of cleats.

Separation of Conductors. A separation of at least ¼″ should be maintained between the supporting screws or nails and the conductor. Cleats for voltage up to 300 volts must separate the conductors 2½″ from each other and ½″ from the surface wired over. Conductors that are No. 8 AWG or larger, when supported on solid knobs, Fig. 3, should be secured by wires having the same type of insulation as the conductors.

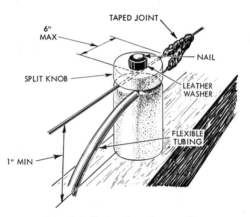

Fig. 4. Split knob and tap splice

A split porcelain knob, Fig. 4, is used for conductors of No. 10 AWG gage or smaller. It is provided with a groove on each side of the nail so two wires of the same polarity and circuit can be attached. Where a tap is made from the main wire of a circuit, both conductors must be supported within 6″ of the tap, Fig. 4.

Nails for mounting knobs should be at least the 10-penny size. Screws should be long enough to extend into the wood at least one-half the height of the knob.

The correct method of making a right-angle turn with two conductors is shown in Fig. 5. On the left, solid porcelain knobs are used, and on the right, cleats are used. It is important that the

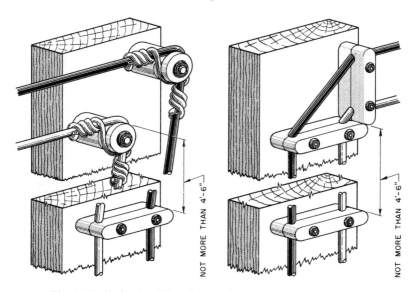

Fig. 5. Methods of making right angle turn with two conductors

2½″ clearance between conductors be observed on turns as well as on straight runs.

Protection on Ceiling, Side Walls, and Floors

On low ceilings, guard strips should be installed, Fig. 6. These must be not less than ⅞″ in thickness and at least as high as the insulators.

Protection on side walls must extend not less than 7′ from the

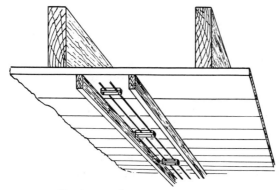

Fig. 6. Guard strips on low ceiling

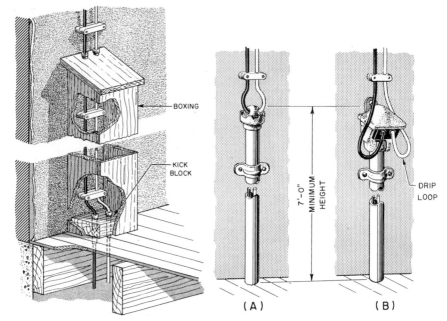

Fig. 7. Wood boxing for side
wall protection

Fig. 8. Pipe or conduit used
for side wall protection

floor. It must consist of a substantial boxing closed at the top, Fig. 7, and must provide an air space of 1″ around the conductors. The wires, it should be noted, pass through bushed holes. When porcelain knobs are used instead of cleats, the wires must be at least 3″ apart. Wires passing through floors should be properly bushed, and a kick block should be used to guard tubes at the floor line.

Protection of Wires. Conductors may be protected from injury by the use of pipe or conduit. When wires pass inside the pipe, each wire should be inclosed in a separate nonmetallic flexible tubing, called *loom*. The loom must be in one piece and must extend from the porcelain support of the wires at the bottom to their porcelain support at the top.

When conduit is used for protection of the wires, a terminal fitting having a separate bushed hole for each conductor, Fig. 8A, is used. The same fitting is attached to the bottom end of the conduit (not shown). This method is also used when the open-wiring system is changed to conduit. Wires in the pipe or conduit must not contain splices.

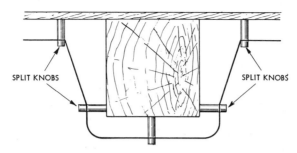

Fig. 9. Method of breaking around beams with split knobs

In wet locations a service head is attached to the upper end of the conduit, Fig. 8B. The wires should be arranged to form a drip loop so that water cannot enter the conduit. Conductors that form the drip loop should be at least 2″ away from iron parts of the conduit and service head so that water will drip free of the pipe.

Support for Wires. Wires require rigid support even under ordinary conditions. On a flat surface, supports should be provided at least every 4½′. If wires are likely to be disturbed, the distance between supports must be shortened.

A long board is nailed between widely-spaced beams, and the insulators are mounted on the board. The method of breaking around beams with split knobs is illustrated in Fig. 9. In buildings of mill construction, cables of not less than No. 8 AWG (where not likely to be disturbed) may be separated about 6″ and run from

Fig. 10. Cleat type lampholder
Courtesy of Bryant Electric Co.

Fig. 11. Cleat type lampholder
Courtesy of Pass and Seymour

Fig. 12. Surface type snap switch
adaptable to open knob & tube wiring
Courtesy of Bryant Electric Co.

Fig. 13. Ceiling pull-switch
adaptable to knob-and-tube wiring
Courtesy of Bryant Electric Co.

timber to timber without breaking around beams. Drop cords should
be attached to the circuit wires through a cleat, Fig. 10. The best
installation of wall sockets and switches is with special cleat-work
fittings, Figs. 11, 12, and 13. Other fittings, when used, should be
mounted on small porcelain knobs. Conductors should be dead-
ended on a cleat or on a knob, as shown in Fig. 14.

Precautions Against Dampness and Acid Fumes

The rules for open work in buildings subject to moisture or acid
fumes are somewhat different as regards insulator supports. Porce-

Fig. 14. Dead-ending a wire at a knob

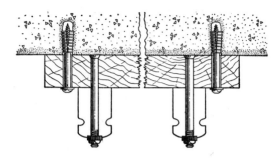

Fig. 15. Wood strip independently fastened by bolts and
expansion shells for supporting porcelain insulators in damp places

lain or glass knobs must be used. When installed on brick, concrete, tile, or plastered walls or ceilings, these insulators should be attached to wooden or metal strips or blocks, fastened independently by means of expansion or toggle bolts, Fig. 15. Nails or screws driven into wooden plugs are not permitted by the National Electrical Code.

Wires should have standard rubber-insulating covering for protection against water and against corrosive vapors. Weatherproof or rubber-insulated conductors should be separated at least an inch from the surface wired over. Sockets should be of weatherproof type, Fig. 16, and drop lights should be hung on No. 14 AWG standard rubber-covered wire.

Cutouts and switches should be mounted on small knobs in iron cabinets.

Stringing Wires

One method of stringing wires is to fasten each conductor to a

Fig. 16. Weatherproof socket

Courtesy of Pass and Seymour

double support at one end, then pull it across joists to the other end of the run. After a terminal cleat or knob is installed, required insulators are attached to joists along the way. If the wires are small, one or two turns around the handle of a hammer or bar, as shown in Fig. 17A, will provide good leverage. An easy method for tightening small wires is to incline the end knob "backward," as in Fig. 17B. When the split knob is driven up, with the wire in the knob, the conductor is tightened automatically as the knob comes to an upright position.

(A)　　　　　　　　　　　　　(B)

Fig. 17. Method of taking up slack in small conductors

CONCEALED KNOB-AND-TUBE WIRING

Where Used. Concealed knob-and-tube work, like wiring on insulators, it not permitted in congested districts of large cities. It is sometimes employed, however, in residential areas, and in small towns. The electrician should learn fundamental principles underlying this method, because there are thousands of knob-and-tube installations already existing which must be repaired and extended from time to time.

Essential Requirements. The NEC prohibits the use of this type of wiring in commercial garages and other hazardous areas. Only single conductors may be run on knobs. Conductors must be supported at intervals not exceeding 4½′, separated at least 3″ from each other, and maintained at a distance not less than 1″ from the surface wired over. At distribution centers and other points where a 3″ separation cannot be maintained, each conductor must be encased in a continuous length of flexible tubing.

Where practicable, conductors shall be run on separate timbers or studding. They shall be kept at least 2″ from piping or other conducting material, or separated therefrom by a suitable non-conductor such as a porcelain tube. In accessible attics, conductors must be protected by running boards or guard strips if within 7' of the floor, unless run along sides of joists or rafters. Conductors passing through cross timbers in plastered partitions must be protected by an additional "mud" tube that extends at least 3" above the timber. The more important points in knob-and-tube installation are illustrated in Fig. 18.

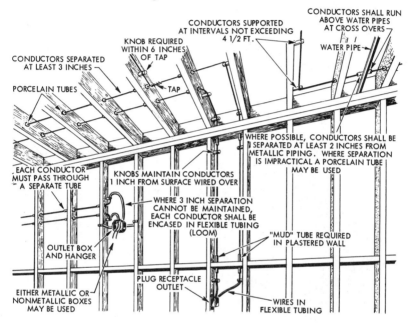

Fig. 18. Typical knob-and-tube installation

Outlet Boxes. The standard metal outlet and switch boxes may be used with knob-and-tube wiring. It is permissible, however, to use a porcelain, Bakelite, or composition box for this purpose. A number of these wiring accessories are now available, and give service comparable to the metal outlet box. We must remember that in a knob-and-tube job it is possible for the hot wire to be grounded to the side of a metal box and not be noticed until touched by someone in contact with ground. For safety reasons, composition outlet, switch, and receptacle boxes may be desirable.

Insulated boxes, Fig. 19, are constructed of plastic or composition materials. Their dimensions are comparable to those of metal boxes. Insulated outlet box covers and plaster rings are also available.

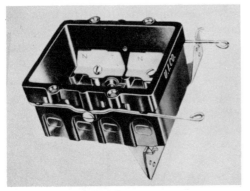

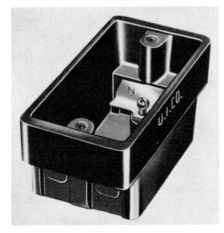

Fig. 19. Insulated outlets, switch boxes and covers

Courtesy of Union Insulating Company

When metal boxes are used, flexible tubing must extend into the box. It is secured there by clamps or other means. When insulated boxes are used, the tubing need extend only to the outer surface of the box.

The method for installing an outlet box between joists is shown in Fig. 20. It is important that the box extend down far enough to come flush with the ceiling finish. For lath and plaster, allow about 1″; for composition board, allow about ½″; and for masonite, allow about ½″ or ⅜″, depending on the thickness selected.

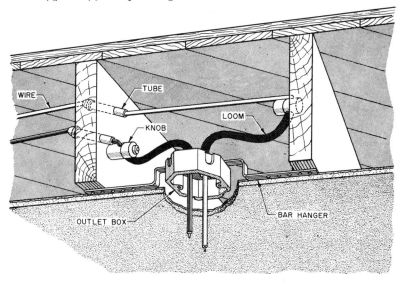

Fig. 20. Mounting outlet box using bar hanger

METHOD OF WIRING NEW BUILDINGS

Since this form of wiring is not widely used on new construction, the installing procedure will be outlined only briefly. Certain details which are applicable to all forms of wiring, including knob-and-tube, will be investigated in the next chapter in connection with cable wiring.

The first step is to mark locations of outlets with chalk or crayon, using standard symbols to indicate switches, plug receptacles, and lighting outlets. Ceiling lights are noted on the floor at the approximate centers of rooms. Locations of the fuse panel and of the service equipment are also marked. Outlet boxes are now fixed in place and the fuse panel set.

Circuits will be discussed in a later chapter. Here, we assume that the number and distribution of circuits has already been determined. Holes are bored to accommodate the various runs, using a brace and bit, a boring machine, or, more commonly, an electric drill with a bit extension.

Wires are pulled through the holes, and porcelain tubes slipped over the conductors as they are drawn from joist to joist or stud to stud. Knobs are installed where needed, and conductors drawn tight. Wires are skinned, and joints made up, after which they are soldered and taped. Short lengths of flexible tubing are then installed, where necessary, at the various outlets and at the fuse panel. At the same time, the ends of flexible tubing are secured in the boxes. When feeder and service runs are completed, the electrician's work is done until carpenters, painters, and other mechanics have finished their jobs.

METHOD OF WIRING OLD BUILDINGS

In wiring old buildings where it is not possible to take up flooring, nonmetallic, flexible tubing may be "fished" (shoved through) in partitions or between the floor and ceiling. Only one wire is permitted in each piece of tubing, which must be continuous from outlet to outlet or from knob to knob. This tubing may not be run across notched beams or joists, as is done with armored cables, but must be fished through existing channels in partitions or over ceilings. By removing baseboards, fishing parallel with the floor beams, going through floor plates with porcelain tubes, and picking up the circuits in the attic and the basement, a safe and acceptable installation can be put in with flexible tubing at about two-thirds the cost of an armored cable installation.

REVIEW QUESTIONS

1. What is the oldest type of electrical wiring?
2. What additional step is necessary where wires pass through cross timbers in a plastered partition?
3. How much separation between wires is provided by standard cleats?
4. What type of knob-and-tube wiring would you use for a barn?
5. What is the largest size wire that may be supported by a split knob?
6. What type of insulation is required for tie wires used with solid knobs?
7. How far should wires be separated from gas pipes?

8. State a simple way for tightening small wires in knob-and-tube work.

9. What clearance from the surface wired over is provided by knobs?

10. Can metallic outlet boxes be used with concealed knob-and-tube wiring?

11. What is the purpose of flexible loom?

12. What type of thermal insulation may be used with knob-and-tube wiring?

13. How much clearance from the surface wires over is provided by standard cleats?

14. State the minimum permissible clearance between wires in concealed knob-and-tube wiring.

15. What is the maximum permissible distance between supports in concealed knob-and-tube wiring?

16. May nonmetallic, flexible tubing be fished on new work?

17. What device is used to support ceiling outlet boxes?

18. Are holes for a new knob-and-tube installation usually bored with a brace and bit?

19. What is the first step in beginning a knob-and-tube installation on new construction work?

Chapter ⑦

Methods of Wiring

Nonmetallic Sheathed Cable and Flexible Armored Cable

QUESTIONS THIS CHAPTER WILL ANSWER

1. Under what conditions is Type NMC nonmetallic sheathed cable permitted?
2. What is meant by the term "polarity grouping"?
3. What is the procedure for installing cable wiring on new construction?
4. What materials may be used for underplaster extensions?
5. How are surface metal raceways installed?

NONMETALLIC SHEATHED CABLE WIRING

Types. Nonmetallic sheathed cable consists of two or three insulated conductors, either with or without an additional bare conductor for grounding purposes.

Figure 1 shows the construction of *Dutrax*, one of the brands of nonmetallic sheathed cable now on the market. There are various makes, such as *Braidex, Triex, Romex,* and others. *Romex* was the first on the market, and many in the trade still refer to any make of this kind of material as *Romex*.

The NEC recognizes two types of nonmetallic sheathed cable in No. 14 AWG to No. 4 inclusive: Type NM and Type NMC. The

Fig. 1. Dutrax nonmetallic sheathed cable
Courtesy of Anaconda Wire & Cable Company

outer covering on Type NM is flame-retardant and moisture-resistant. The insulation on Type NMC, besides being similar to that on Type NM, offers additional protection against fungus and corrosion. Both of these cables are available with an uninsulated conductor which is used for grounding purposes only.

Application. Type NM cable may be used for both exposed and concealed work in normally dry locations. It may be installed in the voids of concrete block or tile walls if not subject to excessive moisture or dampness. It cannot be used where corrosive vapors or fumes are present, and cannot be imbedded in concrete, cement, or plaster.

Type NMC cable may be used in locations specifically denied to Type NM above, and it may be run in the shallow chase of a masonry wall that is covered with plaster. If within 2" of the finished surface, however, it must be protected by a ¾" steel plate which is not less than 1/16" thick. Neither of these types of nonmetallic sheathed cable may be used for service entrance, nor may they be installed in commercial garages or other hazardous areas.

Essential Requirements. Conductors must be supported at intervals not exceeding 4½', and within 12" of outlet boxes, cabinets, or fittings. This rule does not apply, of course, to fished work in existing buildings. Cable runs must be continuous from outlet to outlet, as no splice is permitted in the cable itself. It is necessary, therefore, to make all circuit splices in outlet boxes, junction boxes, switch boxes, or other outlet points.

Where the cable is run through holes bored in wooden members, the holes must be near the centers of joists and studs, or at least 2" from the nearest edge. Cable may also be laid in shallow notches cut in wooden members if it is protected at these points against the driving of nails. A 1/16" steel plate is sufficient for this purpose. Where cables are run at angles to joists in unfinished basements, assemblies not smaller than two No. 6 AWG or three No. 8 AWG conductors may be attached to the underside of joists. Smaller assemblies shall be carried through bored holes or placed on running boards. In accessible attics, cables must be protected in the same manner as knob-and-tube installations.

It should be noted that outlet devices made of insulating material can be installed without boxes on exposed work, or where cable is fished in existing buildings. Nonmetallic boxes may be employed with this cable. Figure 2 illustrates the most important requirements connected with the use of nonmetallic sheathed cable.

Method of Wiring New Buildings. The initial steps are identical with those for concealed knob-and-tube wiring. Outlet locations are marked, boxes and fuse panel fastened in place, and holes bored where necessary for the running of circuit wires. Boxes are fastened in place, cable is pulled through holes or stapled to wooden members, as the case may be, and secured in the boxes where this is

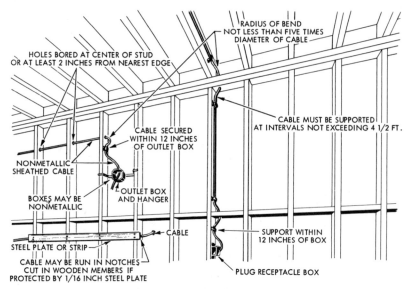

Fig. 2. Typical nonmetallic sheathed cable installation

required. Cable need not be secured to nonmetallic boxes, and it need not be fastened to metallic boxes if supported within 8″ of them. Splices are made, joints taped, circuits run into the fuse panel. Feeder and service conductors are installed.

Procedures for locating ceiling outlets will now be explained. Figure 3 represents a plan view of a bedroom which requires a center light. There are two basic methods: One is to mark the center point on the floor, and then transfer this point to the ceiling by means of a plumb line. The other is to do this work at the ceiling line. Since the latter approach involves taking measurements from a step ladder, the former one is usually more convenient.

The center of the floor area can be found in two ways. Chalk lines may be drawn between diagonal corners of the room, the crossing point being the desired location. The second choice is to measure between opposite walls, finding the spot by trial and error. The latter method would seem more time consuming, but if there are three or more rooms of identical dimensions, distances found for one room may be marked on a light stick and transferred quite rapidly to the others.

So that a straight run of holes may be drilled in ceiling joists, chalk lines are often drawn across the joists to mark a path. With experience, however, the electrician soon learns to dispense with these preliminary steps.

TRIAL AND ERROR MEASUREMENTS

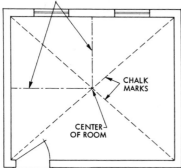

CHALK MARKS

CENTER OF ROOM

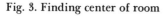

Fig. 3. Finding center of room   Fig. 4. Staples

Although wire may be taken from a coil for long runs in knob-and-tube work by grasping the inner turn and drawing it out, cable may not be unwound in this manner. The reason lies in the fact that cables would twist and become hard to manage. Unrolling is necessary, and a wire reel is very useful for the purpose.

Because of the nature of the material, runs of nonmetalliic sheathed cable (also armored cable) must be cut at each outlet point, and a new run started from there. As a result, more splices are required than in knob-and-tube or in conduit work, since straight-through wires, too, must be spliced.

The outer sheath of nonmetallic cable can be stripped off with a knife, or a cable ripper. In some cases the sheath is removed by a self-contained rip cord provided underneath the braid of certain makes of cable. Then, the wires are skinned in the same manner as ordinary single conductor.

Nonmetallic sheathed cable is fastened to wooden members by means of staples, such as in Fig. 4; it is attached to metallic boxes by cable clamps which are included in the box, or by means of separate outlet box connectors.

FLEXIBLE ARMORED CABLE

Types. The NEC recognizes four types of flexible armored cable: AC, ACT, ACL, and ACV. These cables are distinguished by the

Fig. 5. Type AC armored cable
Courtesy of Crescent Insulated Wire & Cable Co.

covering on the wires: Type AC has rubber insulation, Type ACT has thermoplastic insulation, Type ACL has rubber insulation and a lead sheath, and Type ACV has varnished cambric insulation.

Armored cable was introduced under the trade name of BX by one manufacturer, and is still referred to as such, regardless of the brand. Figure 5 shows the construction of Type AC cable. It can be obtained with either one, two, three, or four conductors. Single-conductor cable is used mostly as a grounding wire.

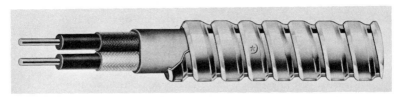

Fig. 6. Lead covered armored cable type ACL
Courtesy of Crescent Insulated Wire & Cable Co.

In order to provide flexibility, the steel covering on BX was not wound so tightly as in older types of cable. This "loose" construction introduced a comparatively high resistance into the steel covering so that it did not provide acceptable grounding continuity. Manufacturers, therefore, placed a copper bonding wire, in sizes No. 14 and No. 12 AWG, underneath the armor and in intimate contact with it. The NEC now provides that Types AC and ACT shall have a copper or aluminum bonding strip throughout their entire length. It is worth noting the correct way of designating wire size and number of conductors in cable. Cables having two No. 14 wires are called "fourteen-two," written as 14/2; those with four No. 12 wires are called "twelve-four," written as 12/4; those with three No. 10 wires are called "ten-three," written as 10/3, and so on throughout the various combinations.

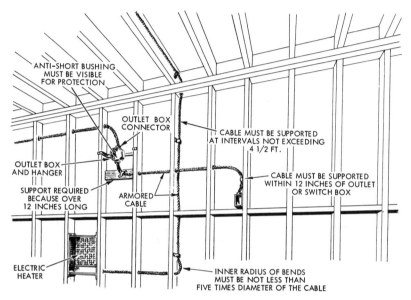

Fig. 7. Typical armored cable installation

Applications. In dry locations, Types AC and ACT cables may be used for either exposed work or for concealed work. They may be fished in the air voids of masonry walls where the walls are not exposed to excessive moisture. They may be used also for under-plaster extensions.

Type ACL cable, Fig. 6, may be imbedded in concrete or masonry, run underground, or used where gasoline or oil is present. No type of armored cable may be used in motion-picture studios, theaters (except for very limited exceptions noted in NEC), hazardous locations, or places where harmful agencies such as corrosive vapors, may be present.

Essential Requirements. Figure 7 shows that the same general rules apply to installation of armored cable as to nonmetallic sheathed cable. Distance between supports must not exceed 4½′, and the cable must be supported within 12″ of an outlet box. In accessible attics this type of cable must be protected in the same manner as nonmetallic sheathed cable.

Connectors. Cables are secured to boxes by clamps which are an integral part of the box, Fig. 8, or by means of separate

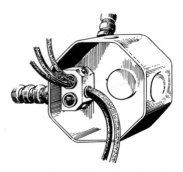

Fig. 8. Octagon box with armored cable clamps

Fig. 9. Outlet box connectors

connectors, Fig. 9. The connectors have a standard conduit thread and a locknut at one end, so that they may be attached to standard boxes. The clamp at the opposite end of the connector body is made in various forms, some of them applicable to both nonmetallic sheathed cable and armored cable.

Polarity Grouping. All wires of an alternating-current circuit, when encased in conduit or in armor, must be within the same enclosure. If not so installed, inductive heating may result.

Preparing the Cable. Armored cable may be cut readily with a hacksaw. The best method is to cut through one of the convolutions about 6″ from the end, and then break it off by twisting back and forth. The outer wrapping may then be removed and the anti-short bushing installed as in Fig. 10.

The connector or clamp can now be attached, and secured to the box. Cable may be fastened to wooden members by staples, similar to those for nonmetallic sheathed cable, or by armored cable straps. To avoid twisting and kinking, armored cable must be taken from the coil by unrolling or by unwinding it from a cable reel.

Fig. 10. Installing antishort bushing

Courtesy of National Electric Products Corp.

Service Entrance Cable—Feeder and Branch-Circuit Cable

Types of Service Entrance Cable. The NEC lists three kinds of service entrance cable: Type ASE, which includes a metallic tape for mechanical protection; Type SE, which is the same as ASE except that the metallic braid is omitted; and USE, which is a special type. Service entrance cable is employed in lieu of wire and conduit for connecting the power company service drops to the

Fig. 11. Unarmored service entrance cable

Courtesy of General Electric Co.

service disconnecting means for the building. Figure 11 shows one kind of service entrance cable.

Applications of Service Entrance Cable. Types ASE and SE are used for overhead services, while Type USE is for underground. Power companies often recommend Type ASE because the metallic wrapping renders it safe from current theft.

Service entrance cable may be employed for interior wiring systems only when all circuit conductors, including the neutral, are of the rubber-covered or the thermoplastic type. Service entrance cable with uninsulated grounding conductor may be used to supply electric ranges, built-in cooking devices, and clothes dryers, or it may serve as a feeder from the service cabinet to another building, provided that the cable has a final nonmetallic outer covering and the alternating current supply circuit does not exceed 150 volts to ground.

Underground Feeder and Branch-Circuit Cable. The NEC lists Type UF cable as acceptable for underground feeders and branch circuit uses. The outer braid of this cable is suitable for direct burial in the ground, being fungus- and corrosion-resistant. Where single-conductor cables are used, all wires of the particular circuit, including the neutral wire, shall be run together in the same

Fig. 12. Armored cable
Courtesy of Crescent Insulated Wire & Cable Co.

trench or raceway. This cable may be employed for interior wiring systems in dry, wet, or corrosive locations. It may not be used for service entrance, nor in commercial garages or other potentially hazardous locations.

Extensions

Underplaster Extensions. Underplaster extensions are permitted only in buildings of fire-resistive construction. Rigid or flexible conduit, armored cable, or electrical metallic tubing may be used

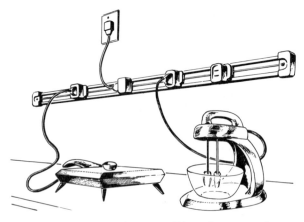

Fig. 13. One form of nonmetallic surface extension
Courtesy of I-T-E Circuit Breaker Co., Bulldog Electric Products Div.

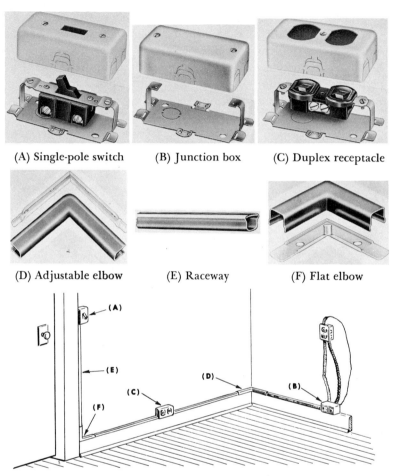

(A) Single-pole switch   (B) Junction box   (C) Duplex receptacle

(D) Adjustable elbow   (E) Raceway   (F) Flat elbow

Fig. 14. Wiremold installation and fittings

Courtesy of The Wiremold Company

for this purpose. Standard sizes of material are required except that for a single conductor, conduit or tubing not less than 5/16″ inside diameter is permissible. A single-conductor armored cable is also acceptable because conductors of a circuit installed as an under-plaster extension need not be contained in the same raceway. Figure 12 shows a type of armored cable which lends itself to such work. These extensions must be confined to the single floor where they originate unless standard sizes of cable and conduit are employed.

Nonmetallic Surface Extensions. This material, shown in Fig. 13, is accepted by the NEC for exposed extensions run from existing outlets in dry residential or office locations. It may not be installed in unfinished basements, attics, or roof spaces, and it is also not permitted where the voltage between conductors exceeds 150, or where corrosive vapors exist.

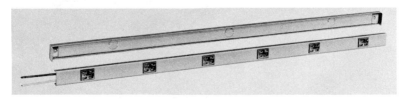

Fig. 15. Multi-outlet assembly
Courtesy of The Wiremold Company

The extension run must not come into contact with metal. It must not be run on the floor or within 2″ thereof. It must be run in unbroken lengths without splices, and may not pass through a floor or partition, or extend outside the room where it originates. The material must be secured in place at intervals not exceeding 8″, except that where connection to the supplying outlet is made with a cord and plug, the nearest fastening shall be not more than 12″ from the plug.

Surface Metal Raceways. Surface metal raceways such as that shown in Fig. 14 are permitted only in dry locations. They should not be exposed to severe physical damage, used in hazardous locations, or subject to corrosive vapors. The largest size of conductor allowed in this form of raceway is No. 6 AWG. The raceway may be extended through dry partitions or floors if the lengths which pass through are unbroken.

Multi-outlet assembly, Fig. 15, is subject, in general, to the same restrictions and allowances as those applicable to standard surface metal raceway. In addition, the code also lists two precautions: First, when run through a partition, no outlet of the strip may fall within the partition. Second, removal of the cap or cover on exposed portions shall not be impeded by the fact that the run extends through the partition. That is, the cover for a portion of the assembly which is outside the partition shall not be locked in by the partition itself.

REVIEW QUESTIONS

1. Can Type NM cable be used for exposed work?
2. Can Type NMC cable be imbedded in concrete?
3. Is it permissible to fish Type NM cable in an interior concrete block wall?
4. Is it permissible to use Type NMC cable in a commercial garage?
5. What type of nonmetallic cable would you install in a dairy barn?
6. Where installed in holes bored in timber, how far should NM cable be from the nearest edge of the joist or stud?
7. What should be the maximum distance between staples where Type NMC cable is fastened to the side of a joist?
8. Must polarity grouping be observed in runs of Type NMC cable?
9. Armored cable must be secured, in general, within what distance of an outlet box?
10. How must the cut ends of armored cable be protected when installed in an outlet box?
11. What is the shortest permissible radius for bends in armored cable?
12. How is grounding continuity assured in Type AC cable?
13. Is armored cable permitted in fished work?
14. Are anti-short bushings required with Type ACL cable?
15. Can service entrance cable with a bare neutral conductor be used for interior wiring?
16. Can Type ASE cable be employed for an underground service?
17. May underplaster extensions be used in frame buildings?
18. Can nonmetallic surface raceway extend through dry partitions?
19. Can surface metal raceway be run through a floor?
20. Can multi-outlet assembly be run through a dry partition?

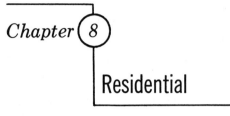

Chapter 8

Residential
Wiring Processes

QUESTIONS THIS CHAPTER WILL ANSWER

1. What are section drawings?

2. What type of auger bit is used on new construction work?

3. What is the procedure for mounting a new switch outlet box in an old sheetrock wall?

4. How would you pass a wall obstruction in an old house?

5. How can you take up floor boards with the least possible damage?

NEW HOUSES

Sectional Drawings

General Inclusions. In addition to the plan view illustrated in Chapter 1, the draftsman or architect also furnishes section drawings of the structure, Fig. 1. They show story height and ceiling height of each floor in the building. *Story height* is the distance from the top of the finished floor of one story to the top of the finished floor in the next upper or lower story. *Ceiling height* is the distance from the top of the finished floor of one story to the finish line of the ceiling of the same story.

Section drawings show what floors are composed of; for example, whether they have wooden joists, Fig. 1A, or some other type of construction, Fig. 1B. This fact influences the method of doing electrical work and the kind of materials needed. Thus, if floors and partitions are wood, concealed wiring may be done with knob-and-tube, nonmetallic sheathed cable, or armored cable—provided local rules allow. If the structure is concrete, either conduit or electrical

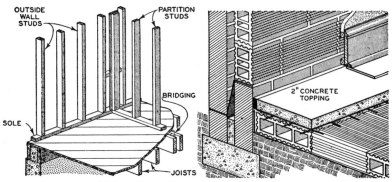

Fig. 1. Section drawings of a frame and a concrete structure

metallic tubing must be employed for this work.

Ceiling Finish. The sectional drawings also show the amount of space there is to be between the bottom of the joists and the finish line of the plaster or other ceiling material, called *ceiling finish* for short, as well as the space between the top of the floor joists and the top surface of the finished floor, called *floor finish* for short. These spaces at times are considerable, and must be allowed for in the selection of the outlet box and the plaster ring or cover, as well as in the placement of the outlet box. Otherwise the edge will either project beyond the ceiling finish or be too far withdrawn from it.

In buildings of wood joist construction, the laths are nailed directly to the bottom of the ceiling joists, Fig. 2. Thickness of lath and plaster combined is approximately an inch, being slightly more for a first-class three-coat plaster job. In some cases, wallboard,

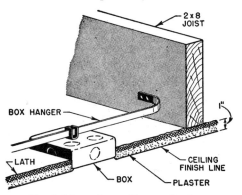

Fig. 2. Mounting ceiling outlet box

sheetrock, or wood may be used for ceiling finish. The materials vary in thickness; hence it will be necessary to check on this point if the work is to be concealed. With a proper hanger it is possible to use a $1\frac{1}{2}''$x4'' box without plaster ring. If a 4'' opening is too large for the fixture canopy a plaster ring may be installed on the box, thus reducing the size of the opening to $3\frac{1}{2}''$.

There are times when the ceiling is furred down several inches. When the amount of furring is greater than can be taken care of

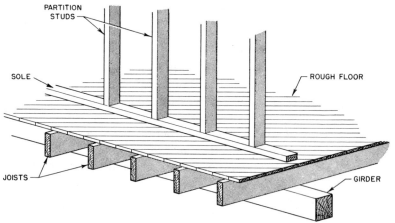

Fig. 3. Partition walls of a frame building

with an extension box, it is necessary to choose between (1) running the conduit in the floor construction above and extending the conduit down to the outlet box, or (2) running the conduit in the space between floor and suspended ceiling.

Floors. There are several types of floor construction. The cheapest method is where the finish floor is laid directly on the joists. Lumber customarily used for this purpose is $\frac{7}{8}''$. In the next cheaper form of construction a $\frac{7}{8}''$ rough floor is laid on the joists then furring strips of the same thickness, the top floor being nailed to the latter. Here an allowance of $2\frac{1}{2}''$ or more is necessary when taking measurements depending on the thickness of flooring lumber. Beyond this there are other types of floors having sound proofing or fireproofing, with or without cinder fill, and wood *sleepers* to which the finished floor is nailed. The allowance for this type of floor may be 5'' or more. Marble and tile floors vary in thickness according to the particular design.

Mounting Heights. Heights for switches and plug receptacles, given in the plans, are from finished floor to the center of the outlet, unless otherwise noted. Proper allowance must be made, when roughing in, for the kind and thickness of finished floor. Where the height given for plug receptacle outlets is low, failure to provide for floor thickness may result in the baseboard overlapping the outlet. The same thing can happen from neglecting to check width of the baseboard which is to be installed. Commonly specified heights for wall outlets are 18″ for plug receptacles (convenience outlets) and 52″ for switch outlets.

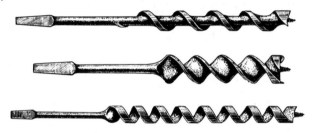

Fig. 4. *(top)* Single-spiral, single-cutter bit; *(center)* Double-spiral, double-cutter bit; *(bottom)* Single-spur car-bit or ship-auger

Construction. There are numerous types of residential designs, requiring different kinds of building materials, such as stone, brick, concrete, cement block, tile block, and wood. In most cases, the interior of the building is of frame construction with floor joists, rough floor, and 2 x 4 studs for partition walls, as shown in Fig. 3.

Floor plans should be consulted to determine the swing of particular doors so that switch outlets may be mounted in the proper locations. A wall switch is always mounted on the striking side of a door. With double doors, the outlet can be installed on whichever side is desired, but must be far enough over so that the switch will not be hidden from view when the door is open. Also, with respect to switch outlets, it is well to check door trim on the plans to make sure that trim will not overlap the switch plate. In general, there should be about 2″ clearance between the edge of the switch plate and the door trim.

Boring Holes

Joists. When joists are bored with a joist-boring machine, holes can be nearly horizontal, thus making the pulling-in of cables somewhat easier than when holes are aslant. The more common method today, however, is to use an electric drill with a bit extension for drilling joists and studs. Holes drilled in this manner are, for a given size of cable, bored one or two sizes larger than would be required for horizontal holes.

Choosing Auger Bits. For new construction work, a bit with a coarse pitch of thread on the screw point for fast feeding and a single cutter head for easy cutting are preferred. See Fig. 4.

The single-spur car-bit (or ship-auger) is best for boring holes in old house wiring. Nails are often struck in this kind of work because there is little choice in the matter of hole locations. This style of bit is not damaged so easily as other types, and it is somewhat easier to put in shape again after striking a nail.

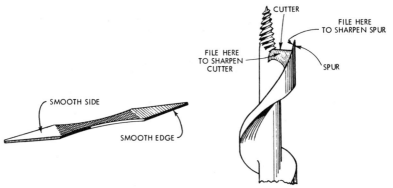

Fig. 5. Auger bit file and enlarged detail of cutter

Sharpening Auger Bits. The single-cutter head is sturdier than the double-cutter head, and it is also easier to file after striking a nail or other hard substance. For this type of filing, use a small file with a fine cut. Figure 5 shows a file made expressly for this purpose. On one end it has teeth in the flat sides and none in the flat edges, and on the other end it has teeth in the edges and none in the sides. These file sides and edges without teeth are called *safe side* and *safe edge* respectively.

File the outlining spur on the inside only, as shown in Fig. 5, because the spur must be cut at the point of maximum diameter for easy boring. The cutter, that is, the edge at the end of the spiral which cuts and lifts the chip, should be filed mostly on its upper or lifting face, Fig. 5, although a bad nick or bulge in the lower face should be filed out. The *lead* of the spur, that is, the distance which the spur extends beyond the edge of the cutter, should never be less than the thickness of the chip. The original angles of the cutting edges should be maintained as closely as possible. Otherwise, the bit will not cut well.

OLD HOUSES

Preliminary Measures

Testing Circuits. It is a good rule to assume that all circuits are electrically alive, or "hot," until they have been checked. And the best method for checking is with a pocket device such as the voltage tester shown in Fig. 6..

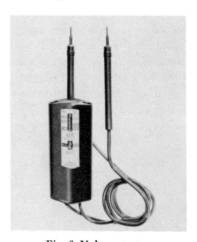

Fig. 6. Voltage tester

Courtesy of Ideal Industries, Inc.

The white wire should be checked, where present, to make sure that it is the neutral conductor. Connecting the tester between this wire and a grounded surface, such as a water pipe, should give no indication of potential. If the building has been wired in recent years, the NEC color code has probably been followed. For a three-wire circuit, the standard coding is: black, white, and red. If the circuit is 115-230 volts, the tester will show 115 volts between the white wire and either the black or the red wire, 230 volts between the black and the red wire, and 115 volts between either black or red wire and a grounded surface.

An inexpensive test set for checking circuits for continuity is shown in Fig. 7. It consists of two dry cells in series with a buzzer. The push button may or may not be included, because one of the

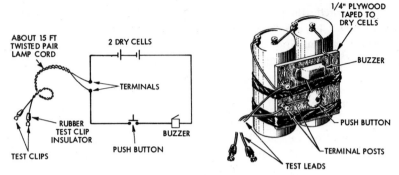

Fig. 7. Illustrating how to build a test set

test clips may be attached to a circuit wire, and the other clip may be used to make or break contact as desired.

Locating the Outlets. This class of work frequently is called *old-house wiring,* or *old work.* It covers the installation of wiring with a minimum of damage to or removal of finish. As the cables must be fished from outlet to outlet, this type of wiring is often a two-man job.

Fig. 8. Rotary or ceiling hacksaw
Courtesy of Misener Mfg. Co.

The first step is to spot the outlets in ceilings and walls, after consulting the owner or his representative. Ceiling outlets are usually placed in the center of the room. In the choice of location for wall brackets, discretion and a knowledge of building construction are essential, because some walls offer more difficulties than others. For example, it is often impossible to fish cable in the furred space between lath on a brick wall.

The plaster-keys which overhang laths on the inner surface may be so large as to seriously impede fishing of cables. Or, the space between the top of a wall and the roof may be so small that access to a hollow space inside the wall is impossible.

The second step is to cut holes for the outlet boxes. Holes should be only large enough to accept the boxes without forcing. For ceiling outlets the boxes are round, ½″ or ¾″ in depth, depending on the thickness of plaster. These switches or receptacles are 2″ x 3″, of a depth to suit the particular condition. Bracket outlet boxes are

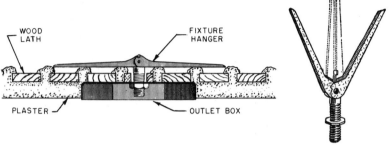

Fig. 9. Ceiling outlet box and fixture hanger

Fig. 10. Old-work bar hanger

either round or rectangular in accordance with the type of canopy or mounting plate used in that particular location.

Figure 8 shows a handy tool for cutting round holes which is used with a brace. Cut the plaster only, the lath being left intact. It will probably be necessary to notch the lath to permit passage of the cables and outlet-box support.

Figure 9 illustrates a ceiling-outlet box and fixture hanger. The view shows the hanger legs extended for supporting the box. They are held in open position by the locknut on the center stem which serves also as a fixture stud. A key through the center of the nipple locks the wings securely in a horizontal position. In another type of hanger, Fig. 10, a stud slides freely along the bar. The whole unit is inserted through the opening, and the stud is drawn along the bar to a central position. A wire, already attached to the stud, is used for this purpose. The locknut is removed and the outlet box, with cables attached, is placed on the stud. The locknut is then set up tight, thus securing box and hanger in place. A hole about

$1\frac{1}{2}''$ in diameter, somewhat larger than for the other type, must be made for this hanger. With either one, the entire weight of the fixture must be carried by the lath. Hangers must be so installed as to bear on as many laths as possible.

Installing Switch Outlets. The process of cutting a hole for a switch outlet is different from that for a lighting outlet because lath

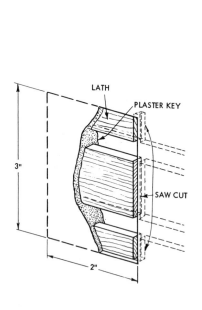

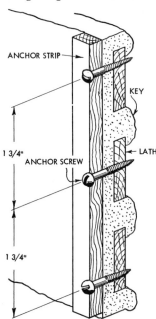

Fig. 11. Wall construction showing location for cutting opening in plaster for switch box

Fig. 12. Wall construction showing method of supporting plaster when cutting opening for outlet

must be cut through as well as plaster. For a standard switch box, the dimensions of the hole should be approximately $3''$ vertical and $2''$ horizontal. Since wood lath is $1\frac{1}{2}''$ wide and the spacing between laths is about $\frac{1}{4}''$, it will be necessary to cut out the middle lath entirely, and to cut off $\frac{1}{2}''$ from each of the adjacent laths.

Dig out enough plaster to locate the lath in the spot where the switch is to be installed. Outline the dimensions of the hole on the wall, using the middle of the exposed lath as a center point. Carefully remove plaster within this outline, as in Fig. 11. Now cut a section $2''$ wide out of the middle lath, and a $\frac{1}{2}'' \times 2''$ strip out of each adjoining lath, using a narrow, fine-toothed saw. The saw should be manipulated gently, especially after the first cut has been made, because supporting studs may be some distance away, and the spring of the lath tends to crack plaster.

Where old plaster is too brittle to sustain even the most careful sawing, a wooden anchor strip may be employed. The strip may be from 1″ to 1½″ wide, ½″ or more thick, and 8″ to 10″ long. Bore three holes through the strip large enough for the screws to pass through. The strip is installed as shown, Fig. 12, with No. 6 or No. 8

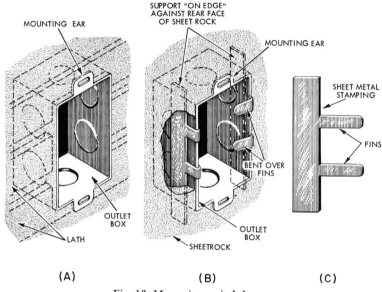

(A)   (B)   (C)

Fig. 13. Mounting switch boxes

wood screws. Pilot holes should be drilled through the laths to prevent splitting. Saw the laths, and transfer the strip to the other side of the opening. If any screw holes fall outside the span of the switch plate when the job is finished, they may be filled with Plaster of Paris, and tinted if necessary.

With sheetrock or with patented lath, such as button-board, the operation is much simpler. The hole may be cut with a thin saw, or even a sharp knife, at any desired place, except directly over a wall stud.

Attachment of a switch box to wood lath or other wall coverings is also an easy task. Fig. 13A shows how a standard switch box is fastened to wood lath, two small wood screws securing the upper mounting ear to the lath against which it rests, and a second pair of screws holding the lower ear in position.

Figure 13B illustrates the method of using patented sheet metal box supports, and Fig. 13C, that of holding a switch box in a sheetrock wall. One of the supports is placed in the hole, standing "on edge" against the back face of the sheetrock, its two projecting fins sticking out through the opening. The fins are bent over, along the wall, to hold the support temporarily while the other one is installed. The switch box is now placed in the aperture, and the fins are bent into the box and down along the inner face, as shown in Fig. 13B. After the second set of fins have been handled in the same way, the box is securely retained.

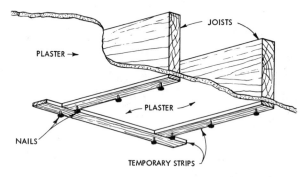

Fig. 14. Temporary strips for supporting plaster when cutting a scuttle hole

Plug Receptacle Outlets. The procedure for mounting plug receptacle boxes on plastered or composition walls is exactly the same as for switch boxes. A more common method, however, is to cut them into the wooden baseboard. After the location is determined, and an outline drawn on the baseboard, four small corner holes are drilled so that a keyhole saw may cut along the outline from hole to hole.

Scuttle Hole. The job of fishing cables is simple when there is enough attic space in a one-story building or, as the case may be, on the top floor of a multi-story building. Usually, there is a scuttle hole for access to this space. If not, it is easy to make one in a closet or storeroom.

Locate the floor joists, and remove plaster between two joists over the approximate area of the desired opening. If wood lath is encountered, nail temporary wooden strips to the joists, as in Fig. 14, driving the nails only part way in. The strips should be about 26" long. Nail end strips at right angles to them to mark the outline of the hole, and saw out the required amount of lath. If the ceiling is sheetrock or patented lath, the strips may be dispensed with, sawing the required distance along the joists, and then at right angles to them until the piece of material is cut free.

The strips may then be removed and a permanent frame installed, as in Fig. 15. Headers and a cover may be constructed as shown.

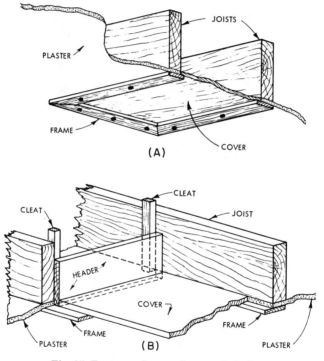

Fig. 15. Frame and cover for scuttle hole

Fishing. Fishing methods are the same, regardless of whether the wiring material happens to be loomed wires in a knob-and-tube installation, or one of the cables. When the term "cable" is employed in this discussion, it will be understood that loomed wires are included as well.

Fishing down to a wall outlet from an attic space is not a complicated procedure. The first step is to bore a hole through the partition plate directly above the point where the outlet is to be located. This point can be located through suitable measurements transferred from the lower floor. The plate is a 2x4 timber, often a double one, between the upper ends of the studs and the joists.

Next, drop a sufficient length of sash chain through the hole, and fish it out from underneath with a piece of hooked wire. Attach the cable to it, and draw up enough to reach the outlet box from which the switch loop is to be supplied. Where headroom is insufficient to permit use of a drill for cutting the hole in the plate, a

Fig. 16. Diagonal and horizontal bridging

chisel and hammer may serve the purpose. This difficulty arises, sometimes, when a gable roof slopes down onto the partition at a fairly small angle.

If there is an unfinished basement under the floor that is being wired, wall outlets, especially baseboard plug receptacles, may be fished from there. Bore through the floor plate or sole, of the partition from underneath. If the baseboard has been removed, it may be easier to drill from above. In this case, other wall outlets around the room may be wired by running cable in bored or notched holes in the studding behind the baseboard.

Passing Obstructions. In most frame walls, braces or bridging blocks are inserted between adjacent studs. They consist of short pieces of 2x4 placed at an acute angle, or else straight across. Figure 16 shows braces in two partitions.

Since either type of bridging hampers fishing operations, steps must be taken to pass through or around it. For example, a switch is to be installed near a door, and at a point under a brace whose location has been determined by measuring the fish chain or by carefully sounding the wall with a mallet or other suitable object. First, take off the door stop, which is at least 1½" wide. Bore a hole through the door casing and the stud, or door buck, at a point above the obstruction, and another some distance below it. Then, cut a channel in the door casing from one hole to the other, slanting off the corners where channel and hole meet, to avoid sharp bends in

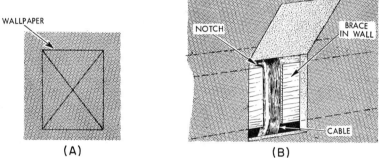

(A)   (B)

Fig. 17. Alternative method of passing wall obstruction

the cable. After the wiring is installed and the door stop replaced, there will be practically no evidence of the work.

If the outlet is to be, say, 3' from the door casing, the same plan may be followed, using one or, if necessary, two bit extensions to reach through the next stud. A door opening may be crossed by taking up the threshold and gouging a channel underneath it. If there is no threshold, a channel will have to be cut at the top of the door opening.

When the outlet location is too far away from a door, another method of passing the obstruction must be employed. Figure 17 illustrates the manner in which bridging is passed during the process of fishing cable from the attic to a plug receptacle in the baseboard. After the bridging has been located with considerable accuracy, as shown by the dotted lines, the outline of a narrow rectangle is drawn with a piece of chalk. If the wall is made of bare plaster, the material is scraped out until the lath is exposed. The lath is cut with a chisel, and a hole gouged in the cross brace. After the cable is installed, the hole is repaired with Plaster of Paris.

If there is wall paper, it may be cut diagonally as in Fig. 17A, or rectangularly as in Fig. 17B, depending on the design of the paper. After proceeding as before, the wall paper is pasted back.

Removing Trim and Floor Boards. When removing a piece of interior trim, use a thin, wide tool to distribute pressure over as large an area as possible. A narrow chisel is apt to leave a mark which is hard to remove or conceal. Start the prying effort at a point above or below the normal line of vision so that marks left by the tool will not be readily observed.

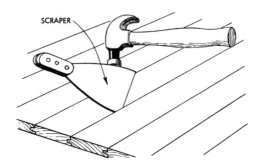

Fig. 18. Illustrating method of cutting tongue in floor board

If cable must be installed at right angles to joists which are inaccessible from below, it is necessary to take up the floor. Begin, if possible, where two pieces of flooring meet, so as to have a loose end available when starting to raise the boards. First, however, the tongue must be cut on each side of this strip of flooring. An ordinary keyhole saw is likely to cut away too much stock from the edges, and cracks will be noticeable when the boards are replaced. A tool which uses a much thinner hacksaw blade, will not cause this trouble. When a hacksaw blade is used without some kind of a holder, it should be inserted so as to cut on the upstroke, because it tends to buckle if made to cut downward.

Before cutting may be started, a hole must be made through the tongue large enough to admit the saw blade. A good tool for the purpose is a painter's scraper, which resembles a putty knife, but which is wider and stiffer. Set one corner of the blade in the crack, and strike sharply with a hammer, as illustrated in Fig. 18.

There may be trouble with flooring nails, especially if they were not driven fully home. They are cut through easily with the hacksaw blade. While flooring is being raised watch carefully for any bulging along the edge of the adjacent strip, and apply pressure there to avoid splintering. Cross bridging runs should be located before removal of floor boards is attempted, in order to prevent the mistake of taking up a section directly above them. Cross bridging in floors consists of "herringbone" braces usually located half way between bearing partitions. They may be spotted from underneath by way of the hole cut in the ceiling for the lighting outlet. After the boards have been raised, the rough underflooring may be cut without difficulty, using a keyhole saw or a short handsaw.

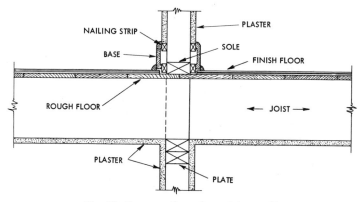

Fig. 19. Construction of partition wall

Removing Baseboard. With this work completed, a method for getting down to the wall switch location on the floor below must be considered. A common device is to remove a section of baseboard, thereby exposing the sole of the partition, usually 2x4 to which the studding is nailed, Fig. 19. Chisel a hole through this timber large enough to admit a hand. Locate the plate of the lower partition, and bore a hole large enough for the cable.

Cutting a Pocket. An alternative plan to removing baseboard is to cut a pocket between two joists, close to the upper wall, and then drill the plate of the lower partition. This method must be used, also, when there is no partition directly above the lower one.

If possible, locate the pocket at the end of a strip of flooring. Sometimes it pays to start at the end of a strip one or two joists beyond the desired location so that it will not be necessary to saw a floor board. Cut the tongue, and begin raising the strip from one end, going as far as possible before running the risk of splitting that portion in which the tongue is still intact.

Insert a bar, chisel, or similar object between the floor and the raised strip, as in Fig. 20. If there is room enough to cut the under-flooring over section A and to perform the other work, nothing further need be done to the floor. Should more working room be needed, the strip must be cut at point B, which is the middle of one of the joists, in order to get support for both ends of the strip when the board is replaced. A thin saw should be used to avoid cutting away too much stock.

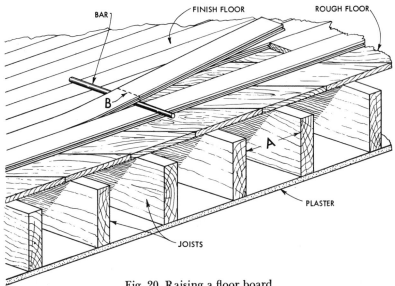

Fig. 20. Raising a floor board

When a pocket is to be cut between two floor joists, say in the middle of the room, proceed as in Fig. 21. Bore a $\frac{1}{4}''$ hole through the strip that is to be cut, the hole being as close to the joist as possible. Repeat the operation at the other joist. Cut the tongue and remove the strip. Before replacing it, nail a substantial cleat against the face of either joist to support the piece of flooring.

Occasionally, time may be saved by cutting a pocket in one joist space and another two or three spaces away, instead of taking up flooring all the way across. The three joists, Fig. 22, can be bored from one pocket, and the cable fished through.

Do not cut a floor that is finished in parquetry or similar material. If cables must traverse such an area at right angles to joists, take off the baseboard along one of the sidewalls, and bore or notch the studding to take the cable.

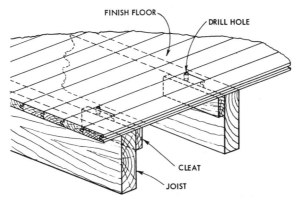

Fig. 21. Floor construction showing method
of supporting floor strip

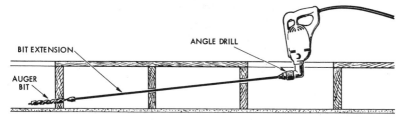

Fig. 22. Boring floor joists with auger-bit, extension and electric angle drill

Replacing Wooden Floor. Rough flooring must be replaced, at least over the joists, as well as paper or other lining that was present, in order to bring the strip to its original level. Nailing must be well done so the floor will not creak when walked on. Use flooring nails at each joist, driving them well into the strip with a nail set. The holes may be closed with putty or plastic wood, stained to match the color of the flooring.

Installing Special Lighting Fixtures

Recessed Fixtures. Recessed fixtures are quite popular today, especially on modernization projects. The code requires that such units shall be mounted so that clearance of the metal enclosure from combustible surfaces shall be not less than $\frac{1}{2}''$, except at the supporting lip.

One type of fixture, which has a built-in junction box protected from direct heat of the lamps, may be connected directly to the circuit wires. With other types, the code demands that circuit wires shall terminate in a junction box not less than 12″ from the fixture, Fig. 23. Wires from the junction box to the unit must be approved for high temperature (usually some form of asbestos-insulated conductor) and must be enclosed in a suitable raceway not less than 4′ and not greater than 6′ long.

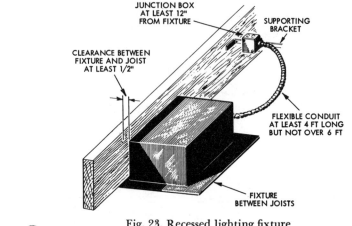

Fig. 23. Recessed lighting fixture

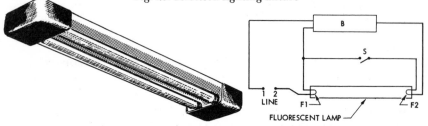

Fig. 24. Fluorescent lamp and basic circuit wiring

Electric Discharge Lighting. Another popular lighting unit is the fluorescent lamp shown in Fig. 24A.

The fluorescent lamp is a glass tube which is coated on the inside with a powder that glows, or fluoresces, when exposed to radiation from an electric arc. It contains mercury vapor and a small quantity of argon gas. An arc is drawn between two filaments, one at either end of the tube.

Figure 24B illustrates one circuit for this type of lamp. When the circuit is connected, current flows from line wire 2 to filament F_1, automatic starter S, filament F_2, ballast B, and line wire 1, completing the circuit. Ballast B consists of a reactance coil with or without a transformer. Starter switch S is normally closed so that its points are touching. When the switch is turned on, current flows through F_1, S, and F_2 in series. As the filaments become hot enough to give off electrons, the contacts of S open, and the reactance coil in ballast B creates a high voltage which sends an electric spark through the tube from F_1 to F_2. The momentary discharge causes the gas to form a conducting arc between the two filaments, and the lamp is in operation. A more recent type dispenses with the starter switch by including a high-voltage transformer in the ballast.

REVIEW QUESTIONS

1. What dimension that is not shown on a plan appears on a section drawing?
2. What is meant by the term story-height?
3. What is a common height for wall switch outlets?
4. At which side of a door is the switch usually mounted?
5. State the NEC color code for a three-wire circuit.
6. State the difference between the plate and the sole.
7. Name a common device for testing disconnected circuits.
8. On which side of the spur must an auger bit be sharpened?
9. Name a common type of patented lath.
10. What is the best tool for cutting the tongue of a floor board?
11. What kind of tool is best for removing interior wooden trim?
12. Must lath be cut when installing a ceiling outlet in an old house?
13. What should be the first step when preparing to install a switch at a selected spot on a sheetrock wall?
14. Are switch boxes fastened in old sheetrock walls by means of wood screws?
15. What is a pocket?
16. Must pockets always be made close to partitions?
17. What structural elements must be replaced before a pocket strip is reinstalled?
18. Is it always necessary to remove floor boards in order to install wires in a direction at right angles to joists?
19. What trouble occurs if the nailing of floor boards is insufficient?
20. Name two comparatively new types of popular lighting units.

Chapter ⑨

Methods of Wiring

Conduit

➤

QUESTIONS THIS CHAPTER WILL ANSWER

1. With respect to physical characteristics and use, what are the differences between rigid conduit, flexible conduit, and electrical metallic tubing?

2. How may conduit be fastened to a tile wall?

3. What is the difference between a rigid conduit hickey and one used for electrical metallic tubing?

4. How would you make a right-angle bend of a given dimension in a piece of thin-wall conduit?

5. How would you make a saddle in a piece of rigid conduit?

Conduit Materials

Types of Conduit. Metallic conduit for general-purpose wiring is made of either steel or aluminum. There are three types: rigid conduit, flexible conduit, and electrical metallic tubing. The latter type is usually referred to as "thin-wall conduit," and the two terms will be employed interchangeably here. Rigid conduit is made to the same dimensions as standard gas pipe, the trade size of which is expressed, according to the nominal inside diameter, as ½", ¾", 1", and so on. Such conduit dimensions are shown in Table I.

Rigid Type. Rigid or non-flexible conduit, Fig. 1A, is supplied in 10-ft. lengths. It is treated to make the interior smooth, and coated outside to protect against corrosion. Steel tubing coated with black enamel is termed *black conduit*. When plated with zinc by one of the galvanizing processes it is termed *galvanized conduit*. Black conduit has very limited application, not being permitted if exposed to moisture or corrosion.

E.M.T. Type. The wall thickness of electrical metallic tubing (EMT), Fig. 1B, is only about 40 percent of that of rigid conduit. It is easier to handle than its rigid counterpart, and may be installed more rapidly because of the type of fittings used with it. The largest obtainable size of electrical metallic tubing is 2″.

TABLE I. CONDUIT DATA, DIMENSIONS, AND WEIGHTS

THIN WALL		TRADE SIZE INCHES	RIGID	
Diameter Inches			Diameter Inches	
Outside	Inside		Outside	Inside
0.706	0.622	½	0.840	0.622
0.922	0.824	¾	1.050	0.824
1.163	1.049	1	1.315	1.049
1.508	1.380	1¼	1.660	1.380
1.738	1.610	1½	1.900	1.610
2.195	2.067	2	2.375	2.067
Not Made		2½	2.875	2.469
		3	3.500	3.068
		3½	4.000	3.548
		4	4.500	4.026
		4½	5.000	4.506

Flexible Type. Flexible metallic conduit, Fig. 1C, was first introduced under the name of Greenfield. It is still referred to by this term throughout the trade. This conduit is put up in coils instead of lengths. It may be used in lieu of the rigid type in most dry locations. The most common application, however, is for flexible connections at motor terminals.

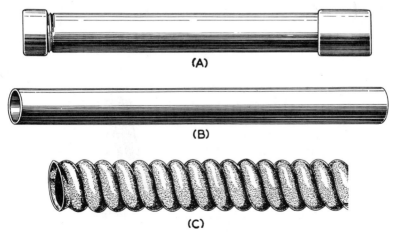

(A)

(B)

(C)

Fig. 1. Conduit

Conduit Elbows. While conduit follows closely the regular dimensions of standard gas pipe, (shown in Table I), there is a marked difference in the bends or elbows, usually called *ells*. Gas-pipe ells are castings with short radius, for 90° or 45° bends, but conduit ells are standard only for 90° bends. They are made from conduit tube bent to a long radius to facilitate the *pulling-in* of the wires. This radius varies from nearly six diameters (inside)

TABLE II. CONDUIT ELBOWS

THIN WALL		TRADE SIZE, INCHES	RIGID	
Radius Inches	Offset Inches		Radius Inches	Offset Inches
5.75	8.625	1	5.75	8.625
7.25	10.000	1¼	7.25	10.000
8.25	11.000	1½	8.25	11.000
9.50	13.625	2	9.50	13.625
		2½	10.50	15.687
		3	13.00	17.750
Not Made		3½	15.00	20.000
		4	16.00	21.312
		4½	18.00	23.500

for 1″ ells to four diameters for 4″ ells. Table II contains dimensional data for both thin-wall and rigid conduit elbows. Note that in Fig. 2 the offset is greater than the radius, because beyond the completion of the 90° arc there is, at each end, an added straight length, roughly equal to 2″ in the smaller sizes, increasing to nearly 5″ in the largest size. This offset must be allowed for in determining the length of conduit which is to be used in a run which includes an ell. Since the offsets are not always the same for ells of a specific conduit size, it is wise to have the ell on the job before taking final measurements. Also, it is advisable to check the ell to make sure that its angle is a true 90°, especially if two or more are to be installed side by side.

The ½″ and ¾″ ells are usually bent on the job from a length of conduit, with either an elbow former, a hickey, or by hand in a pipe vise, and are not listed in Table II. No conduit elbow should be bent to a shorter radius than six times the inside diameter of the conduit.

Thin-wall ells are stocked by electrical supply houses. They are not generally used in sizes smaller than 1″.

In addition to standard ells, bends of longer radii, commonly called *sweeps,* are used especially where multiconductor cables are to be installed. These sweeps are not carried in stock by supply houses, but must be ordered from the factory. They can also be formed on the job by means of a hydraulic bender.

Couplings. Conduit couplings are similar to standard gas-pipe couplings except that they are finished just like conduit. If necessary, gas-pipe couplings can be used.

Bushings and Locknuts. Where a conduit enters a box or other housing through a hole that is not threaded, the junction between

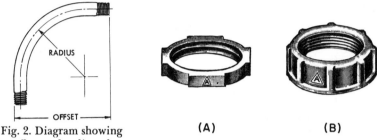

Fig. 2. Diagram showing offset and radius of conduit elbow

(A)   (B)

Fig. 3. Locknut and conduit bushing
Courtesy of Appleton Electric Co.

the two is made firm and secure by means of a locknut, Fig. 3A, and a bushing, Fig. 3B. The locknut is screwed onto the threaded end of the conduit on the outside of the box and the conduit is then passed through the conduit hole in the box, the bushing is then screwed onto the conduit as far as it will go and the locknut set up tight against the side of the box.

Some localities require two locknuts, one inside the box in addition to the bushing, the other on the outside of the box. The National Electrical Code demands this double-locknut construction only where the voltage to ground exceeds 250 volts. When in doubt, consult the local code and the electrical inspector.

Thin-wall Fittings. Thin-wall conduit couplings are of two general types: Those for use anywhere that thin-wall conduit is permitted, commonly called *watertight,* and those permitted in dry places only.

Watertight Couplings. Water tight couplings consist of a short body which slips on the tube, Fig. 4A. Each end carries a male thread, and is split in line with the axis of the bore. A gland screwed onto the threaded end of the body compresses the split end, thus tightly clamping it to the tube.

Couplings for use in dry locations (which include those imbedded in concrete) are usually of the following types.

Setscrew Coupling. The setscrew connector is shown in Fig. 4B, the tightened setscrew forming a slight dent in the tube. The screw being under pressure forms a good metallic ground connection from one tube to another. A coupling has two setscrews, one for each tube.

Indenter Coupling. Indenter type couplings and connectors, Fig. 4C, are slipped over the end of the tube, and a special tool is used to exert tremendous pressure on opposite sides of the connector and tube. If the indenter tool is moved a quarter turn, two more

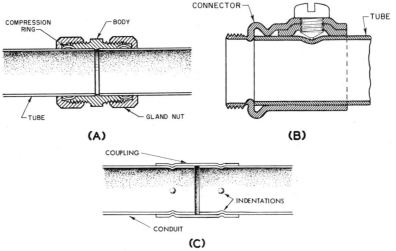

Fig. 4. Couplings for electrical metallic tubing. (A) Watertight coupling (B) Setscrew coupling (C) Indented coupling

indentations are produced thus forming a tight connection between tube and fitting. An illustration of the indenting tool will be shown later in this chapter.

Self-locking Coupling. Self-locking couplings are similar in construction to the watertight ones, Fig. 4A, except that a compression ring is permanently assembled in a non-threaded fitting. When connectors or couplings are tapped onto the end of the electrical metallic tubing the compression ring grips the tube securely.

Connectors. At one end this connector has a body which slips over the thin-wall tubing, at the other end a short male-threaded extension, Fig. 5, for insertion into the outlet box. A locknut screwed onto this thread clamps box and connector together. Connectors are made in compression, setscrew, indenting, self-threading, or self-locking types, the same as couplings.

Adapters are used when it is necessary to secure thin-wall conduit to a fitting which has a threaded hub. The adapter is slipped onto the thin-wall tubing. When the thread is tightened, the device grips the tube and holds it in place.

Fig. 5. Conduit fittings—connectors

Conduit Fittings

Flexible conduit is fastened to outlet boxes by means of connectors similar to those used for armored cables. There is a clamp at one end for gripping the conduit, and a short thread at the other end for attaching to the outlet box by a locknut.

Fig. 6. Square, octagonal, and rectangular outlet boxes

Outlet Boxes

An outlet box or the equivalent must be inserted at every point in the conduit system where access to enclosed wires is necessary. They fall into three general classes, square, octagon, and rectangular, Fig. 6. Each of these comes in various widths, depths, and knockout arrangement. There are knockouts in the sides and bottoms for readily making conduit entries. A knockout is a round indentation, punched into the metal of the box, but left attached by a thin edge or by narrow strips, and forced back into the opening. It can be removed easily with a hammer or pliers. Bushings and locknuts are used to secure this type of box to a run of conduit. Outlet boxes can be obtained in black-enameled or galvanized finish and can be used for either concealed or open work where there

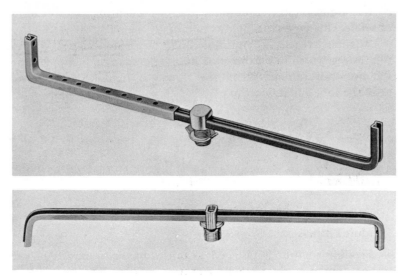

Fig. 7. Two straight-bar hangers

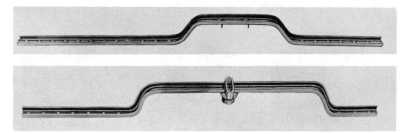

Fig. 8. Two offset-bar hangers

are no volatile, explosive or flammable vapors, dust or similar hazards present, and if not exposed to the weather.

Although there is no hard and fast rule as to the kind of box needed for a specific purpose, the general practice is to use the octagon type for lighting outlets, the other two for switches or receptacles. Outlet boxes are also used for junction and pull boxes, or in place of an ell, when making a 90° turn.

Fixture Studs and Hangers. In addition to provision for conduit entries the octagon or the square outlet box has, at the back, a central knockout for a fixture stud. Fixture studs must be provided where the outlet is intended for a heavy lighting fixture. One type of stud in the boltless type is held in place by a locknut. An older type was fastened with bolts. Two kinds of fixture studs are shown in Figs. 7 and 8—one in combination with a straight-bar hanger, the other with an offset hanger.

Outlet-Box Covers and Extension Rings. Covers for boxes are of numerous forms, each having a particular use. The *blank* type closes the outlet entirely; the *drop-cord cover,* Fig. 9, has a central hole for the passage of lamp cord. The *plaster ring,* also shown in

Fig. 9. Drop cord, plaster ring, and raised switch box covers
Courtesy of Appleton Electric Co.

Fig. 9, is used on wall or ceiling boxes. Octagon boxes do not always require plaster rings; their need is governed by structural conditions on the particular job and the size of fixture canopies which are to be used. Plaster ring covers should be used only where necessary because they hinder the work of installing wires and making splices. Square boxes used for lighting outlets must always have plaster rings. Special rings can be obtained with internal ears drilled and tapped so that switches or other devices can be mounted on them.

The *switch* cover, shown in Fig. 9, is used on square boxes, for switch and receptacle outlets. The raised opening in the cover permits the use of devices having rather deep bodies, and serves also as plaster ground, where needed. These covers can be obtained with various amounts of projection, from $\frac{1}{4}''$ to $\frac{3}{4}''$, to accommodate different thicknesses of wall finish.

Fig. 10. Octagon and square outlet-box extension rings
Courtesy of Appleton Electric Co.

Covers are not made for rectangular boxes because these are intended only for switches and receptacles. Tapped holes allow switches or receptacles to be attached directly to the box.

Extension Rings. Octagon and square outlet box extension rings, shown in Fig. 10, come in a variety of depths from $1\frac{1}{2}''$ up. They are similar to the outlet boxes in every way, except that instead of being closed in the back, they are open, having only a narrow flange to act as a seat when installed on a box. These rings are used to bring the edge of the outlet box out to the face of the plaster in case the box has been mounted too far back, or to provide additional space when an extra deep box is required.

Fig. 11. Toggle bolts with different types of screw heads
Courtesy of Star Expansion Bolt Company

Fastening Devices

There are many devices on the market for securing objects to walls, ceilings or floors of other than wood construction.

Toggle bolts come in two general types, as shown in Fig. 11. In one, the toggle is hinged to the nut; in the other the toggle is hinged to the head of the bolt. The toggle arms are normally held in the outward position by springs within them. These devices can be used on hollow walls, ceilings, or similar locations. The first type, with toggle on the nut, is somewhat handier to use than the other one. To install it, drill or punch a hole large enough to pass the arms. Unscrew the toggle nut. Push the bolt through the hole in the base of the device to be mounted; then screw the nut on the bolt a few threads, so the arms of the toggle will fold back over the bolt. Next, push the toggle through the hole in the wall and screw up on the bolt until it is tight. It may be necessary to pull out on the bolt to hold the toggle arms against the inner surface of the wall, because that is the only way to keep it from turning.

If more than one toggle bolt is used to hold the device, all must be inserted into their holes before the first is tightened. Greater care is required when installing the other type because the bolt may slide back into the wall and become lost.

Anchoring Devices

Where it is desired to mount an object on a dense surface instead of a hollow wall, some form of anchor must be employed. They are designed for application to a variety of surfaces, and experience will teach the best application of each.

Expansion Shells. There are various types of expansion shells. Figure 12 shows the Rawl anchor and calking tool and Fig. 13 shows

Fig. 12. Rawl anchor and calking tool

Courtesy of Rawlplug Company

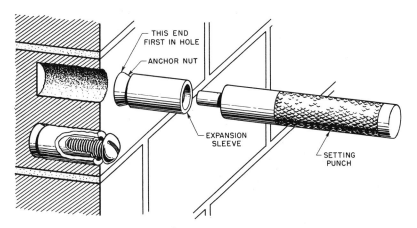

Fig. 13. Method of setting A-J anchors

the Ackerman-Johnson expansion shell. To install, drill a hole of the proper diameter and depth, insert the shell, and expand it by using a tool provided for this purpose, striking two or three moderate blows with a hammer. The smaller sizes are for machine screws only, but larger ones are made for lag screws and machine screws. Their holding power depends on the expansion of a lead sleeve. This expansion is caused by the wedge action of the conical nut, which is drawn into the sleeve by the threads of the screw.

Rawl Plugs. Rawl plugs, Fig. 14, are made from jute, chemically treated and compressed, and are used with wood screws. The screw, the hole, and the plug to be used should all be the same size. The plug must be deep enough in the hole so the unthreaded shank of the screw does not reach it. To install, drill the hole, insert the plug, put in the screw and tighten it. The threads will force the jute fibers into small crevices in the wall of the hole, thereby providing a firm anchorage. The plug should fit the hole snugly. If the hole has been drilled too large, use the next larger size of screw and plug.

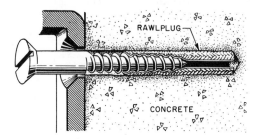

Fig. 14. Illustrating the use of the Rawl plug
Courtesy of Rawlplug Company

Rawl Drives. A one-piece device for making attachments to masonry and kindred substances, is shown in Fig. 15. Known as a Rawldrive, it is installed by drilling a hole of the exact size of the drive, inserting the latter and forcing it home with a hammer. The hole must be deep enough so the end of the drive does not reach bottom. These devices will not hold in crumbly, weak, or yielding substances.

Fig. 15. Construction of Rawl drives
Courtesy of Rawlplug Company

Homemade Anchorage. When the need arises for one or more stud anchorages in masonry and there are no suitable ready-made devices available for the purpose, they can be made as shown in Fig. 16. Obtain the required machine bolts. Drill holes into the masonry large enough to take the heads of the bolts and of such depth that the shoulder of the head is below the surface a distance of 1″ to 2½″ depending on the size of the bolt. Undercut the walls of the holes, and clean out the dust. Insert the head of a bolt, placing it at right angles to the surface of the masonry. Next, fill the hole with molten lead and calk it well. If the masonry is cold, it is a good idea to warm the adjacent area so the lead does not chill before it flows into

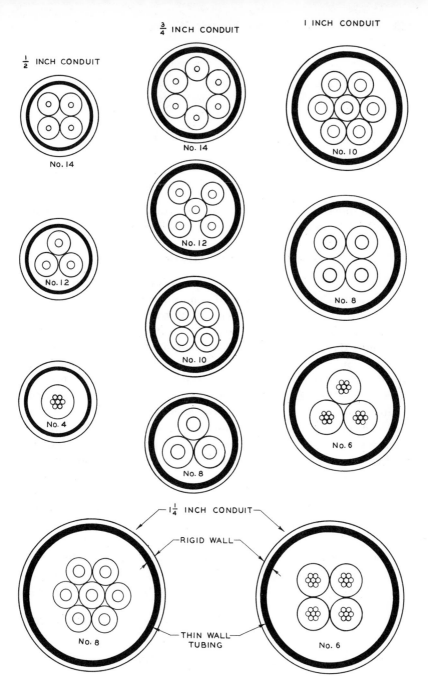

½ INCH CONDUIT

No. 14

No. 12

No. 4

¾ INCH CONDUIT

No. 14

No. 12

No. 10

No. 8

1 INCH CONDUIT

No. 10

No. 8

No. 6

1¼ INCH CONDUIT

RIGID WALL

No. 8

THIN WALL TUBING

No. 6

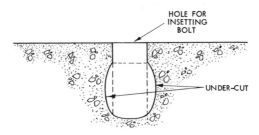

Fig. 16. Home-made anchorage

the small indentations. A washer placed against the head of the bolt before lead is poured will increase the holding power considerably.

WIRE CAPACITY OF CONDUIT

The number of wires permitted by the National Electrical Code is the same for rigid and thin-wall conduit, size for size. It is based on the ratio of the combined cross-sectional area of the wires to the cross-sectional area of the bore. The full-page illustration, page 137, shows the actual size of thin-wall conduit (solid black circle) or of rigid conduit (outer and inner circles), and the maximum number of wires of different sizes which can be installed. With four or more braided wires, the cross-sectional area of all wires cannot be greater than 40 percent of the cross-sectional area of the conduit. For lead-encased cables, the cross-sectional area cannot be greater than 38 percent of the cross sectional area of the conduit when there are four conductors, or 35 percent when there are more than four.

It should be noted that, from the installer's viewpoint, it is preferable to use three single-conductor, lead-encased cables than one 3-conductor. First, the pull is much easier; second, there is not so much waste if terminals to which they are to connect are not close together.

For braided wires, the National Electrical Code permits not more than a total of four 90° bends in a run of conduit between any two boxes. It is better to limit them to three, especially for the larger sizes of wires.

For lead-encased cables, the Code imposes a limit of two 90° bends in one run.

TOOLS AND METHODS

Small Tools. An electrician who does conduit work for an electrical contractor is expected to furnish such small tools as pump pliers,

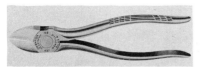

Champion DeArment Tool Company

Crescent Tool Company

Champion DeArment Tool Company

Crescent Tool Company

Champion DeArment Tool Company

Crescent Tool Company

Champion DeArment Tool Company

The Stanley Works

Crescent Tool Company

The Stanley Works

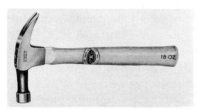

Champion DeArment Tool Company

The Stanley Works

The Stanley Works　　　　*The Stanley Works*

Fig. 17. Electrician's hand tools

Fig. 18. Pipe vise stand and bender

Courtesy of The Nye Tool & Machine Works

Fig. 19. (*Left*) Spiral flute reamer; (*right*) Indenting tool

Courtesy of The Nye Tool & Machine Works

side-cutting pliers, diagonal cutting pliers, knife, pipe wrench, miscellaneous screwdrivers, a hammer, a few chisels, a hacksaw frame, a plumb-bob, keyhole saw, folding rule, steel tape, hand brace, and a crescent wrench or two. All these tools are illustrated in Fig. 17.

The contractor furnishes blades for the hacksaw frame. Those for cutting rigid conduit have 18 to 24 teeth per inch; those for cutting thin-wall conduit have 24 to 32 teeth per inch.

Miscellaneous Shop Tools. The contractor supplies a pipe vise, Fig. 18, and a reamer, Fig. 19 (*left*), for removing burrs from the end of the freshly cut pipe. For use with thin-wall indenter couplings and box connectors, an indenting tool, Fig. 19 (*right*) is required. A knockout cutter, Fig. 20, is also a necessary device.

Dies and Die Stocks. These are contractors' equipment. The earliest type of die was made of a solid piece of metal which was held

Fig. 20. Knockout cutters
Courtesy of Greenlee Tool Company

Fig. 21. Conduit die and adjustable die block
Courtesy of The Nye Tool & Machine Works

in a two-handled stock, Fig. 21 *(left),* and a different die was needed for each pipe size. An improvement soon followed in the shape of adjustable die blocks which could be moved in the stock to fit two or more sizes of pipe, Fig. 21 *(right).*

A later improvement was the ratchet stock with assorted dies, Fig. 22. The dies are easily inserted in or removed from the stocks to accommodate different sizes of pipe within the range of the tool.

The die stock with adjustable guide, Fig. 23, has receding teeth which move in or out to suit various diameters of pipe, through manipulation of a sector handle. This device is used for threading large sizes of conduit.

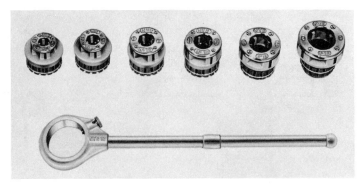

Fig. 22. Ratchet die stock with assorted dies
Courtesy of The Nye Tool & Machine Works

Fig. 23. Die stock with adjustable guide
Courtesy of The Nye Tool & Machine Works

Hickeys and Pipe Benders. The hickey in Fig. 24A is for rigid conduit; the one in Fig. 24B, with the flanged socket, is for thin-wall conduit. The hickey with the curved socket, Fig. 24C, is also for thin-wall conduit, and is the type more commonly employed. On large jobs, where there are numerous duplicate bends or offsets, a bench type of tool, such as Fig. 25 is frequently called upon. The hydraulic bender, Fig. 26, finds application where large sizes of conduit are involved.

How to Make a Right-Angle Bend in Rigid Conduit. Suppose it becomes necessary to make a 90°-bend in a piece of ½″ galvanized rigid conduit, so that it will fit into an outlet box whose lower edge is 10⅜″ from the floor. Add ⅜″ for locknut and bushing inside the box, giving a total length of 10¾″. Mark this distance from the end of the conduit, and measure 5″ in a backward direction to a second mark, Fig. 27A.

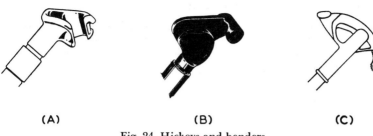

(A)   (B)   (C)

Fig. 24. Hickeys and benders

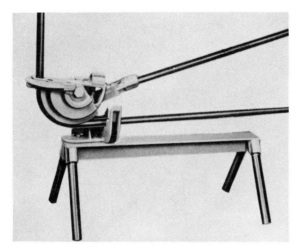

Fig. 25. Bench type conduit bender

Courtesy of Greenlee Tool Company

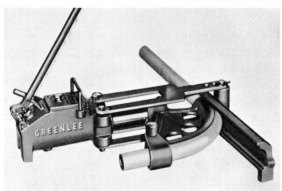

Fig. 26. Hydraulic conduit bender

Courtesy of Greenlee Tool Company

Lay the conduit on the floor, and against a wall or other obstacle to prevent back thrust, then slip the hickey over the end. Rest a foot on the pipe to hold it flat, and adjust the back edge of the hickey to the second mark. Place the other foot on the conduit near the head of the hickey. Press down on the handle, meanwhile, until an angle of about 45° is made, as shown by dotted line *1* in Fig. 27B. Slide the hickey upward a short distance, and make an-

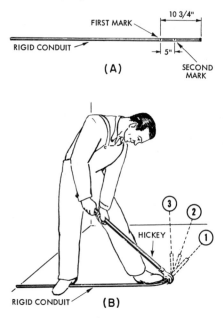

Fig. 27. Bending rigid conduit

other bend of about half the first angle, as indicated by dotted line *2* in the figure. Move the hickey upward about the same amount to finish the bend as shown by dotted line *3*.

Measure the completed bend. If a bit long, slide the hickey down to the neighborhood of its first position, and unbend a small amount. Then, raise the hickey to its final position, and take a rather sharp "bite" to form a right angle. If, on the other hand, the bend is too short, the process is reversed, first making a greater than 90° bend at the original mark, and then taking a backward "bite" to make a right angle. Finally, check to see that the bend is exactly square.

The distance measured in the "backward" direction, stated here as 5″, varies according to size of conduit, material of which it is made, type of hickey, and precise method of performance. The distance may be anywhere from 4½″ to 5½″ for ½″ rigid conduit, and as much as 8″ for 1″ conduit. Actual experience soon makes this operation simple and routine.

How to Make Offsets in Rigid Conduit. Offsets require the exercise of more skill than right-angle bends. The forming of complicated offsets and saddles can be learned only through the trial and error process of experience. Certain principles can be applied, how-

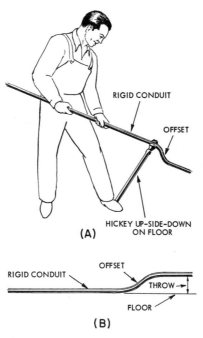

Fig. 28. Making offset in rigid conduit

ever, and varied to suit material, tools, and personal methods of the workman. The general procedure may be explained with the aid of an example.

An offset is to be made in a length of ½″ rigid conduit, to pass along a 5″ wall projection. The distance from the threaded end of the conduit to the beginning of the offset is 12″. Mark this distance on the tubing, and measure backward about 2½″. Stand the hickey upside down on the floor, Fig. 28A, with a foot against the lower end of the handle to steady it. Now insert the conduit into the hickey until the back edge of the hickey rests upon the second mark, and bend a 45° angle. Turn the conduit over, sliding it through the hickey only far enough to clear the curved portion, and bend a 45° angle, so that the short end and the long end are parallel.

Check the offset by holding it on the floor and measuring, Fig. 28B, to determine if the throw is correct. If too great, or if not great enough, proceed as with the right-angle bend to correct the difficulty. If the wall projects, say 10″, rather than 5″, it will be neces-

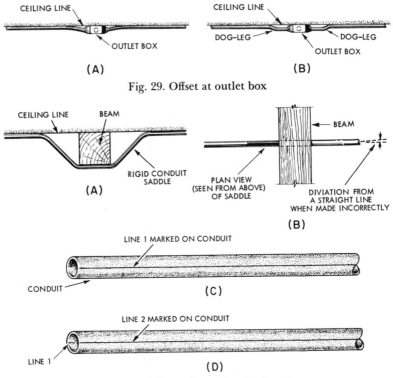

Fig. 29. Offset at outlet box

Fig. 30. Saddles and marking for bending

sary to slide the conduit through the hickey a few inches after completing the first bend, before making the reverse one. Practice soon enables you to make correct distances and allowances for a given size offset.

A useful offset is the dog-leg at outlet boxes in exposed work. Figure 29A shows how conduit looks when not offset at the box. Figure 29B shows the neat appearance when the dog-leg is formed at each side of the box.

Saddles, like that in Fig. 30A, are merely double offsets. One who masters the offset will have no trouble with the saddle. The most common fault in making offsets or saddles is illustrated in Fig. 30B. When the conduit is not turned exactly halfway round after the first bend is made, the reverse bend is out of line with the first one. A good plan for avoiding this difficulty is to mark a line, such as (1) in Fig. 30C, the full length of the piece. Next, rotate the conduit halfway, and scribe line (2) directly opposite line (1), as shown in Fig. 30D. Make the first bend with line (1) uppermost, then make the reverse bend with line (2) uppermost. In this way either bend will be exactly "true" with the other. With enough practice, the turning-over maneuver becomes automatically accurate without the need for guide lines.

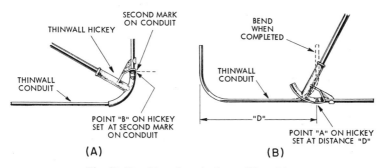

Fig. 31. Bending electrical metallic tubing

How to Make a Right-Angle Bend in Thin-Wall Conduit. Since the bending radii of rigid and thin-wall conduits are the same, the initial measurements are alike. In order to make a $10\frac{3}{4}''$ bend in a length of $\frac{1}{2}''$ thin-wall conduit, a mark is written at a point $5\frac{3}{4}''$ from the end ($10\frac{3}{4}''$ minus $5''$). Point B on the tool is set at this mark, Fig. 31A, and a 90°-bend is made in a single movement of the handle.

For making "U" bends, the head of the bending tool usually has a second mark, A, on the curved arc. The first bend, at the left in Fig. 31B, is fashioned in the usual manner. To make the second one, measure the distance D, and scribe the spot on the conduit. Reverse the tool, setting its point A on the mark, and make a right-angle bend as indicated by the dotted line.

How to Make Offsets and Saddles in Thin-Wall Conduit. Saddles are somewhat easier to make with thin-wall than with rigid conduit. Suppose that an offset with an 8" throw is required. In making the first bend, Fig. 32A, move the handle of the tool until it is in a vertical position, which means that a 45° angle is being produced. Reverse the hickey, as in Fig. 32B, and slide the hickey along the conduit until the throw, as measured along the vertical handle, is equal to the desired amount. Set the end of the handle on the floor and grasp the conduit, bending it until the two legs are parallel.

In order to make this offset into a saddle, simply mark the saddle width on the conduit, and then proceed to make a "reverse" offset of the same dimensions.

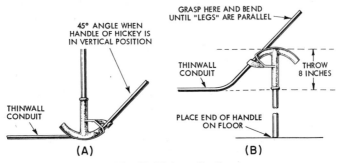

Fig. 32. Thin-wall offset

How to Install Thin-Wall Conduit. The preliminary operations of spotting outlets and determining general circuit layout are the same for a conduit job as for a knob-and-tube or a cable installation. Hence, they need not be repeated here.

At one time the use of rigid conduit in partitions and ceilings was a laborious and time-consuming operation. Thin-wall conduit, largely because of the types of fittings specially adapted to it, makes an easier and far quicker job. Therefore, it has effectually replaced rigid conduit for this kind of work.

Figure 33 shows two methods of running thin-wall conduit in these locations: boring timbers, and notching them. When boring, holes must be drilled at least twice as large as the tubing in order that it may be inserted between the studs. The tubing is cut into rather short lengths, calling for a multiplicity of couplings, and thus increasing the cost of material. Nevertheless, this is the better of the two alternatives.

Because of its weakening effect upon the structure, notching should be resorted to only where absolutely necessary. Notches should be as narrow as possible, and in no case deeper than $\frac{1}{5}$ the stock of a bearing timber. A bearing timber supports floor joists or other weight.

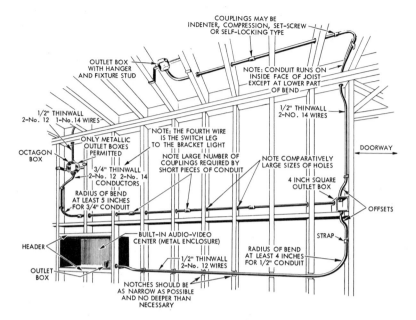

Fig. 33. Installing thin-wall conduit in a frame building

Essential features which must be observed in this kind of work are noted in the illustration.

Fishing. After conduit and outlet boxes are in place and other rough work finished, wires can be pulled in. On short runs, small conductors may be pushed through from outlet to outlet. In most cases, however, it saves time to use a fish tape.

Fish tape, illustrated in Fig. 34A, is made of tempered steel, and is supplied in coils of 50′, 100′, and 200′. To avoid twisting and tangling, it should be kept in a reel. The fish steel may be obtained in thicknesses of .030″, .045″, and .060″. Standard widths are: $\frac{1}{8}$″, $\frac{3}{16}$″, and $\frac{1}{4}$″. The .060″ × $\frac{1}{8}$″ is the most popular size for general wiring.

Formerly, the tape often came with a hook on the end, as shown at Fig. 34C. If it had no hook, one could be readily made with a pair of pliers. Today, some form of leader is usually fastened to the end of the tape. Two such devices are shown in Fig. 34B and Fig. 34D, the smooth ball, and the flexible terminal strip. Either type makes fishing comparatively easy. They have rendered nearly obsolete the older laborious trial-and-error method for getting a fish steel through a difficult run of conduit.

After the tape has been inserted, conductors are attached to the socket of the device and drawn out at the other end. The drawing-out process may be made easier by one of the numerous pulling devices, one of which is illustrated in Fig. 34E.

If the conduit is long, or complicated with bends, so that con-

(A) Fish tape
Ideal Industries Inc.

(B) Fish tape leader
Ideal Industries Inc.

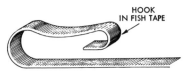

(C) Fish tape with hook

(D) Smooth ball leader
Thomas & Betts Company

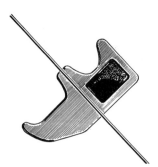

(E) Fish tape puller and method of use

Fig. 34. Fish tape and accessory items

ductors tend to stick, the wires may be coated with talc, soapstone, flax soap, or other non-corrosive substance to lubricate them as they enter the far end of the run.

REVIEW QUESTIONS

1. Name three general types of conduit.
2. What is the largest obtainable size of electrical metallic tubing?
3. Name three kinds of outlet boxes.
4. In case an outlet box is mounted 1½" too far back in a partition, what simple corrective measure is possible?
5. What type of coupling is used on a run of thin-wall conduit that is exposed to the weather?
6. How are indenter couplings fastened to thin-wall conduit?
7. Are self-locking couplings screwed onto thin-wall conduit?
8. What is the most common application for flexible metallic conduit?
9. What is the common name for flexible metallic conduit?
10. Where are toggle bolts used?
11. What are Rawl plugs made of?
12. Name a simple one-piece device for fastening objects to masonry walls.
13. Is thin-wall conduit permitted to carry as many wires as the same size of rigid conduit?
14. How many teeth per inch would be acceptable for a hacksaw blade used in cutting thin-wall conduit?
15. What tool is used to remove burrs from the inside of freshly cut pipe?
16. Can a rigid conduit hickey be used for bending thin-wall conduit?
17. In preparing to make a 10" right-angle bend in a piece of ½" rigid conduit, how far would you measure "backward" to determine the spot at which to place the hickey?
18. What is the most popular size of fish tape?
19. What device may be employed to aid a fish steel to pass around bends in conduit?
20. In general, what are the two methods for installing thin-wall conduit in wooden partitions?

Chapter 10

Large Appliances

Space Heating
Wiring for Motors

QUESTIONS THIS CHAPTER WILL ANSWER

1. What is the difference between a central heating installation and a duct-heating installation?
2. How are ranges and clothes dryers grounded under NEC rules?
3. How are water heaters wired for connection to off-peak meters?
4. How is heating cable installed on the ceiling of a residence?
5. What are the essential considerations in wiring for electric motors?

LARGE APPLIANCES

Electric Ranges. The first large electric appliance to gain popularity in the home was the electric range. Early models had open, wire-wound heating elements that were rather hard to keep clean. Bread crumbs and other food particles collected around the open coils and carbonized there. Ranges with the newer type of enclosed elements, like the range in Fig. 1, are not liable to this objection. They are clean, efficient, and equipped with automatic timing devices that make cooking a comparatively simple chore.

A three-wire, 115-230 volt supply is needed for this unit. Figure 2 shows a simplified diagram of service entrance and circuit wiring for a range installation. The method for determining sizes of circuit and service wires will be taken up in a later chapter devoted to wiring design.

The wiring shown here is installed in either rigid or thin-wall conduit. It could be done with armored cable, or even with service entrance cable, providing local authorities allow it. Under the NEC,

Fig. 1. Modern electric ranges
*Courtesy of General Electric Company and
Hotpoint Div. of General Electric Co.*

service entrance cable may be used for the range circuit as well as for the service itself if the cable has a nonmetallic outer covering. Nonmetallic sheathed cable can be used for the range circuit only, and not for the service.

The range circuit terminates in a heavy-duty, 50-amp receptacle, and the range is equipped with a three-wire cord and plug. The frame is grounded to the neutral conductor of the cord.

The service panel at the left, Fig. 3, is equipped with a main pull-out type disconnect switch and eight plug fuses. The fuses protect eight lighting and utility circuits. If an electric range were to be included in the installation, a second pull-out switch would be needed. It would be found, usually, directly below or to the right of the main pull-out. The NEC permits six separate disconnecting means from the service wires of a single-family occupancy, but local authorities often demand a main disconnect switch as well.

The other panel in Fig. 3 makes use of circuit-breakers instead of fuses. The two large ones at the upper left are the main service disconnecting means; the other two control the range circuit. The

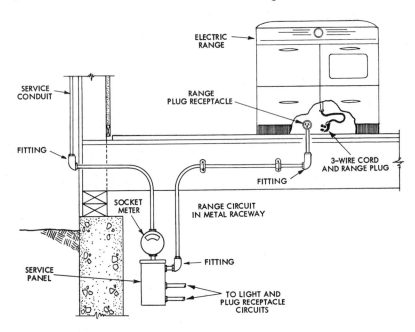

Fig. 2. Service and circuit arrangement for electric range installations

eight smaller circuit-breakers are connected to lighting and utility circuits.

The receptacles, Fig. 4, are for either flush or surface mounting. It is customary to use the flush type in new construction, and the

Fig. 3. Service panels

Courtesy of Federal Electric Co.

Fig. 4. Range plug and receptacles

Courtesy of Pass & Seymour, Inc.

Fig. 5. Modern oven and cooking top

Courtesy of Hotpoint Division of General Electric Co.

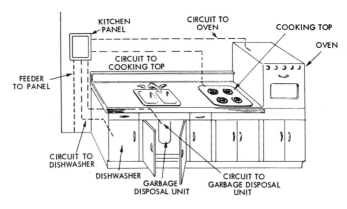

Fig. 6. Circuits to kitchen equipment

surface type on alteration work. The three-wire cord and plug is the same in either case.

Built-In Cooking Units. The electric range of Fig. 1 has been supplanted, to a large extent, by equipment like that in Fig. 5. The oven and the cooking top are like separate parts of a complete range which has been divided for the sake of appearance and ease of operation.

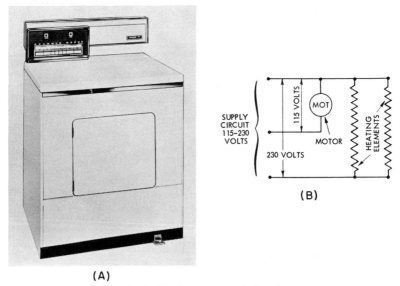

(A)

(B)

Fig. 7. Clothes dryer and circuit
Courtesy of Hotpoint Division of General Electric Co.

Under NEC rules, the two units may be connected to a single 50-amp circuit, or they may be connected to individual circuits of lower rating. A common practice is to run a feeder from the service location to a small panelboard in the kitchen. Here, separate circuits shown in Fig. 6 are run to each heating device and to other fixed appliances, such as garbage disposal and dishwasher.

Dryers. The automatic clothes dryer, Fig. 7A, is another useful item in the home. The average size of the 230-volt unit is 5000 watts (5 kw). A separate circuit is run from the service location or a distribution panel, terminating in a plug and cord connection similar to that for the electric range. The receptacle and plug for the dryer need not be larger than 30-amp. The NEC permits grounding the dryer frame to the neutral conductor of the three-wire cord. A three-wire, 115-230 volt circuit is employed here because the heating elements are designed for 230 volts, while the motor is rated at 115 volts. The heaters are connected between the 230-volt wires, the

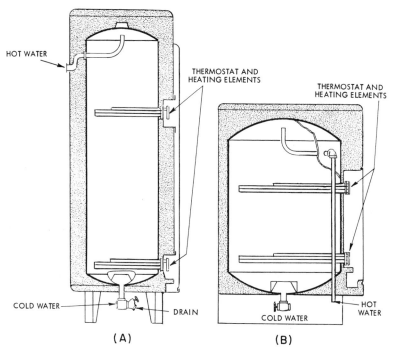

Fig. 8. Round and cabinet type waterheaters
Courtesy of Wesix Electric Heater Co.

motor between one of these wires and the neutral conductor, as shown in Fig. 7B.

Water Heaters. The round water heater, Fig. 8A, has two heating elements. These elements may be connected in a number of ways to suit needs of the user, or to enable him to take advantage of special rates sometimes offered by power companies. In one mode of operation, both heating devices come on at the same time if their thermostats so direct. Another arrangement permits the lower element to draw current only after the upper one has been disconnected from the circuit wires by its self-contained relay. The home type of water heater, however, usually has a single element. The cabinet heater is shown in Fig. 8B.

In some parts of the country power companies give special heating rates, if current is taken only during hours when the supply

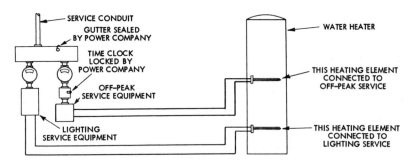

Fig. 9. Off-peak water heater circuit

lines are not heavily loaded. Figure 9 shows a simplified scheme for making use of this plan. One of the two elements of the water heater is connected to the lighting service meter, the other to an off-peak meter and a time clock which permits current to flow during certain prescribed hours. The timeclock is sealed by the power company to prevent tampering with the mechanism. The element connected to the lighting meter is the smaller one, which maintains a desired temperature after the large element has created the proper degree of heat.

SPACE HEATING

Central Heating. A central electrical heating system, Fig. 10A, is similar to a gas furnace. It has a large heating chamber which contains the resistors, and a system of air ducts that lead to wall or

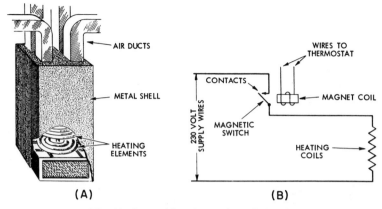

Fig. 10. Central heating unit and circuit

floor outlets in various rooms. Flow of heat is governed by a thermostat (a subject which will be discussed later), and a magnetic switch.

When the switch closes, Fig. 10B, current flows to the heating elements; when it opens, current stops. Some units are designed with two, three, or more elements, each governed by a separate switch or thermostat. The thermostats, in this case, are placed at different locations on the premises.

Duct Heating. A blower circulates air through the ducts, as shown in the illustration, Fig. 11. There is a heater in the main duct, near the blower, and an individual heater in each of the branch

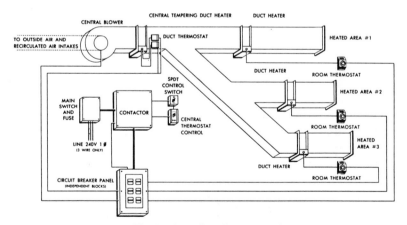

Fig. 11. Duct heating system
Courtesy of Wesix Electric Heater Co.

ducts. A master thermostat, connected to a magnetic switch, determines whether or not current shall be supplied to the circuit-breaker panel. Individual duct heaters are regulated by separate thermostats which are located in the several rooms connected to the branch ducts. Circuit breakers in the panel protect the duct heaters, the wires leading to them, and wires leading to the motor.

Floor Furnaces. This type of furnace, Fig. 12, supplies heat to separate rooms, or acts as an addition to a central heating system

Fig. 12. Floor Furnace
Courtesy of Wesix Electric Heater Co.

when located in hallways, stairwells, or other places which are not adequately heated by the main furnace. Floor furnaces may be controlled automatically by thermostats, or manually by wall switches.

Wall Heaters and Portable Devices. A wall heater may be recessed, as in Fig. 13A, or surface mounted. It offers a combination of radiation and convection heating. The term *radiation* applies to heat rays sent out at right angles to hot surfaces. The term *convection* refers to a circulation of air created in the room by action of the heater. Cool air near the floor rises through the heated interior of the unit to pass along the ceiling to the far wall, then downward to the floor, and along it to the heater again.

In addition to the wire-wound elements shown here, wall heating elements are made in the form of glass or ceramic panels which have high-resistance conductors embedded in the material of which they are made. In any case, they serve a like purpose.

Wall heaters are often supplied by individual circuits. Their output is controlled through built-in thermostats, built-in switches, or wall switches. The portable unit, Fig. 13B, operates in the same manner as the wall heater. It is equipped with a cord, and has a

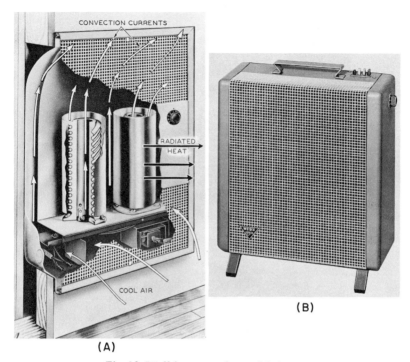

CONVECTION CURRENTS

RADIATED
HEAT

COOL AIR

(A)

(B)

Fig. 13. Wall heater and portable heater

Courtesy of Wesix Electric Heater Co.

plug that can be inserted in a standard plug receptacle.

Baseboard Heaters. Baseboard heating strips may be either recessed on new construction or surface-mounted on existing buildings. These strips can be employed as the sole heating source, or as an addition to a central plant. They are especially valuable in rooms which have large window areas, Fig. 14A. Heat furnished by such units effectively counteracts a tendency for chilling drafts to sweep down over glass panes and across the floor.

Figure 14B shows terminals and wiring space provided for connection of circuit wires, together with the magnetic switch that controls flow of current to the elements. A thermostat mounted on the wall of the room governs the magnetic switch.

The particular unit illustrated here consumes 157 watts per linear foot (approximately 0.7 amp at 230 volts). The degree of heat in the connection box is so low that standard Type R circuit wires may be used.

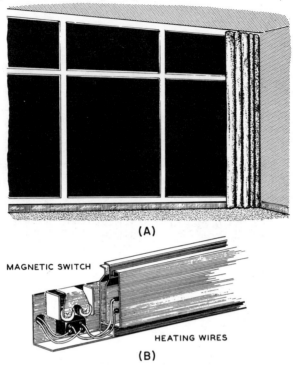

(A)

MAGNETIC SWITCH

HEATING WIRES

(B)

Fig. 14. Baseboard heating

Heated Ceilings. Ceilings may be heated with plasterboard panels that have resistance wires embedded in them. There is a terminal box mounted on the upper face of each 4' x 8' panel. An accessible attic space is required in order that connections may be made between boxes and circuit conductors.

Heating cables are applied to fire-resistant surfaces. They are secured by tape or small staples not over 16" apart, and then covered with a thermally non-insulating plaster. The electrician installs the cables, as shown in Fig. 15, and the plasterer finishes the job.

Heating cables are supplied in factory lengths which must not be cut. They are equipped with non-heating leads at least 7' long, and color-coded red for 230 volts, or yellow for 115 volts. The leads may be run down a partition to a control switch or thermostat. The NEC requires that heating cable shall not run closer than

Fig. 15. Installing cable in ceiling

8″ to a lighting fixture. It may not be installed in closets or over cabinets which extend to the ceiling. Single runs of cable may carry across partitions if they are embedded in plaster.

This type of cable is used also for soil heating in greenhouses, pipe heating where freezing temperatures prevail, deicing of roofs and driveways, or for heating floors. In the latter application, it is permissible to embed the cable in poured concrete.

Ceiling Heaters. Ceiling heaters, Fig. 16, are often used in bathrooms. In general, two types are available: one having a resistance element, the other infrared lamps. The resistance type is usually combined with a Circline fluorescent lamp or with an incandescent lamp, to provide light and heat at the same time. The infrared unit, which also performs both functions, utilizes one, two, or three infrared lamps. It is common practice to connect these heaters to the general lighting circuit.

Resistance heater   Infra-red heater
Courtesy of Nutone, Inc.   *Courtesy of Emerson Electric Co.*

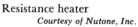

Fig. 16. Ceiling heaters

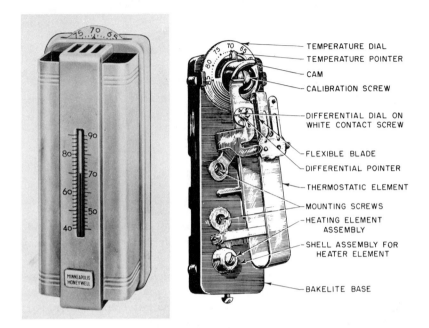

Fig. 17. Outside view of thermostat *(left)* and cutaway view *(right)*
Courtesy of Minneapolis-Honeywell Regulator Co.

Thermostats. Thermostats are manufactured in diverse forms, a precision unit being illustrated in Fig. 17. Contact points are built to handle low-voltage or line-voltage current, as the case may be. Some are designed to be used for either.

Two circuits are illustrated in Fig. 18. In Fig. 18A, the thermostat turns the heating unit on and off. In Fig. 18B, it causes a magnetic switch to open and close. The switch is for turning on

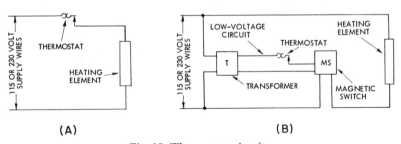

Fig. 18. Thermostat circuits

heating elements or for disconnecting them. A transformer is employed in this illustration to reduce the voltage for the thermostat. Magnetic switches are used in conjunction with line-voltage units, such as this one in Fig. 18A, if contact points are not heavy enough to carry load current. It may be noted that thermostats are employed also to operate electrical valves in a gas-heating system.

Location of a thermostat is an important consideration in any type of heating. It should be placed where it can not be affected unduly by heat from the source. For example, a thermostat used in connection with a baseboard heating installation should not be installed directly above the strip. If possible, it should not be positioned where it is exposed to drafts, as between two doors. The thermostat should be situated within the moderate temperature range, that is, not too high on the wall or too close to the floor. The preferable height is 5'-6".

In new installations, wiring the thermostat is usually a simple task; but in old work, certain procedures must be followed in order to make a neat installation. Figure 19 shows how to drill holes for thermostat wires. After deciding on the location, remove the molding directly below this point, and drill a small "marker" hole through the floor at an angle of about 45°. The molding can be replaced at once. Then drill vertically upward through the basement

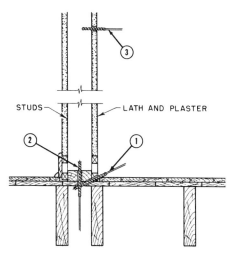

STUDS   LATH AND PLASTER

Fig. 19. Method of installing thermostat wires

ceiling, using the first hole as a guide. Drill a ½" hole at the exact spot where the thermostat is to set. If bridging is encountered, it is by-passed, as explained earlier. The control wires are now fished in, allowing about 6" for making connections.

Plug the hole around the wires with rock wool, or some other insulating material, to prevent drafts of air at the back of the thermostat. If this precaution is not taken, faulty operation may result. When mounting the thermostat on the wall, use a spirit level or a plumb bob to insure a vertical position. The calibration does not hold true unless the unit is perfectly upright.

Line-Voltage Control Units

If the thermostat is a line-voltage type (one connected to the lighting circuit without a transformer), the method of locating and pulling in the wires will be the same; but the line wires will have to be in armored or nonmetallic sheathed cable and secured to an

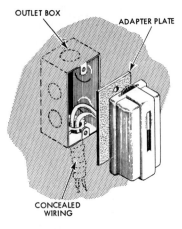
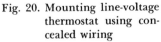
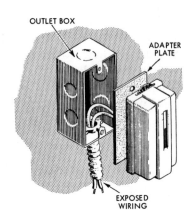

Fig. 20. Mounting line-voltage thermostat using concealed wiring

Fig. 21. Mounting line-voltage thermostat using exposed wiring

outlet box, as shown in Fig. 20. The wiring methods must be the same as for lighting outlets and receptacles in that particular house.

When armored cable is to be run exposed to a line-voltage thermostat, mount the outlet box and thermostat, as shown in Fig. 21. Regardless of the automatic control with which thermostats will be used, they should, of course, be installed as explained. If a line-voltage type thermostat is used, the transformer can be dispensed with and a junction box substituted for convenience in making connections.

UNIT HEATERS—HEAT PUMPS—AIR CONDITIONING

Unit heaters, suspended from the ceiling, are employed mostly in offices. The one shown in Fig. 22 is a combination of gas-fired element and electric motor, operation of the gas valve and motor being controlled by a wall thermostat. The heating element can be a steam coil or an electrical resistor.

Heat pumps are a combination of refrigerating coils and an electric motor. They depend on the principle of transferring heat from one point to another. The motor drives a compressor. In winter, liquid from the compressor is forced through a valve into expansion coils which are usually buried in the ground. As the liquid expands and becomes a gas it draws heat out of the ground. This gas is forced through the compressor to the smaller coils inside the house, at which point it gives up its heat. In summer, the process is reversed, heat being drawn from inside the house and dissipated in the ground. In this case, the pump serves as an air conditioner.

Air conditioning is closely associated with the matter of electric

Fig. 22. Unit heater

Courtesy of Utility Appliance Corp.

heat. The standard air conditioner which is often mounted in a window, cools air which flows through refrigerating coils, and then passes into the room. Most air-conditioning problems are concerned with the understanding of motor wiring. This subject will be taken up later in this chapter.

Practical Considerations. The essential purpose of space heating is to replace heat lost through transmission from inside the premises to the outside. Determination of the amount of power required, expressed in watts, is a complicated process. It depends upon the difference in temperature between that desired inside the house and the prevailing temperature of the atmosphere. It requires a knowledge of average and maximum temperature variations likely to be encountered, and an understanding of heat-transmission constants of building materials. After these factors have been properly assessed, calculations may be performed to arrive at an approximate answer.

Heating engineers achieve more accurate results in new buildings whose exact design features are readily available to them. On existing structures, however, even the engineers run into trouble. Since there is too much risk of creating a dissatisfied customer, the matter of laying out a heating system is not a proper one for the average electrician or contractor.

After considerable experience with the heating of a particular type of structure, one may acquire a fair degree of skill. Until then, experts should be expected to assume risks connected with design of installations. It is within the province of the electrician, nevertheless, to recommend and to install the various types of auxiliary heating devices treated here.

WIRING FOR MOTORS

In order to comply with provisions of the Code, the wireman should understand the basic components of a motor circuit. Its four essential elements, regardless of whether d-c, single-phase, or three-phase, are: (1) size of circuit disconnect switch, (2) size of branch circuit overcurrent protective device, (3) size of motor overcurrent protective device, and (4) size of the circuit conductors. The wiring process can best be illustrated by way of actual examples.

Connecting a ¼ HP, 115-Volt, Single-Phase Motor

The motor in Fig. 23A, rated at 5.8 amps, drives a gas furnace

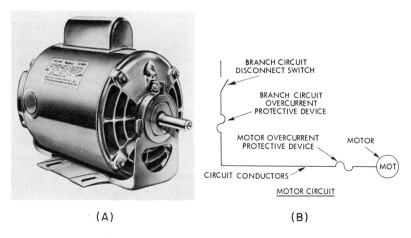

(A)   (B)

Fig. 23. Small motor and circuit diagram

Courtesy of Wagner Electric Corp.

blower in the basement of a residence. A schematic diagram applicable to all motor installations is shown in Fig. 23B.

The maximum rating of a branch-circuit protective device for a motor which is started without series resistors or voltage-reducing apparatus is 300 percent of nameplate current. The nameplate current is 5.8 amps, so that the largest permissible circuit protector is equal to 3 x 5.8 amps or 17.4 amps.

The NEC permits the next larger standard overcurrent device to be used, so that a 20-amp fuse will be acceptable here. The circuit disconnect switch must be able to carry the 17.4 amp starting current, the nearest standard switch being of 30-amp rating. A portable motor of 1 hp or less is not required to have motor overcurrent protection (commonly termed *running protection*), unless it is automatically started. In the present instance, the motor is fixed, and it is automatically started, so that running protection is required for two reasons.

The maximum size of the running protective device can not exceed 125 percent of nameplate current rating for a 40-degree motor, or 115 percent for other motors. The term "40-degree" means that the motor windings, under normal operating conditions, never reach a temperature more than 40° (Centigrade) higher than that of the surrounding air. The term "other motors" includes those rated: "50-degrees," "Short-time Duty," and "Intermittent." Markings of this general nature are stamped on the nameplates.

When running protection is necessary, the rating of the device cannot be greater than 125 percent of motor current rating. Here, the maximum value is equal to 1.25 x 5.8 amps, or 7.25 amps. The motor nameplate should be checked to see if there is a built-in thermal protective device. If so, additional running protection is not required, because the thermal switch in the motor cuts off flow of current when motor windings become too hot. On some motors, current flow will not be resumed until the windings have cooled. On others, there is a button, usually red in color, projecting from one of the endbells. After a shut down from overheating, the motor cannot operate until this button is pressed.

The size of wire needed to supply the motor must now be determined. The NEC provides that the circuit conductors shall have a continuous current rating not less than 125 percent of motor nameplate current rating. This value has already been found as 7.25 amps in connection with the rating of the overcurrent unit. According to NEC Table 310-12 (see Appendix), the allowable current-carrying capacity of AWG No. 14 copper conductor is 15 amps. Since No. 14 is the smallest permissible size of conductor for general wiring, it must be used here.

Finally, the motor should be grounded. If supplied by metal-clad wiring, that is conduit or armored cable, the Code requires that it be grounded. If the wiring is nonmetallic sheathed cable, grounding may be omitted in some instances. It is best at all times, however, to ground the unit. Metal-clad wiring provides its own grounding medium. In the case of nonmetallic installations, where the cable does not have a separate grounding wire, a piece of No. 14 AWG copper wire, either bare or insulated, may be run from the motor to a water pipe.

Connecting a 1½ HP, 115-Volt, Single-Phase Motor

The motor in Fig. 24A is rated at 20 amps. The schematic diagram, Fig. 24B, is similar to that for the ¼-hp motor except that a magnetic starting switch and a push-button station are included. Circuit requirements are caluculated in the same manner as before.

The disconnect switch must have a capacity of $3 \times 20 = 60$ amps, and the rating of the branch circuit protective device must not exceed this value. Motor running protection cannot be larger

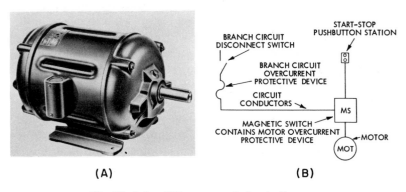

Fig. 24. A 1½ HP motor and circuit diagram
Courtesy of Wagner Electric Corp.

than $1.25 \times 20 = 25$ amps. According to NEC Table 310-12, the nearest size of Type R copper conductor is No. 10 AWG.

Magnetic switches are usually rated in horsepower, the one needed here being marked 1 to 2 hp, single-phase, 115 volts. It must have sufficient capacity to handle the starting current and the continuous running current of the motor. The push-button station has

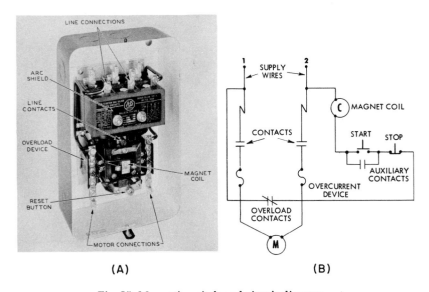

Fig. 25. Magnetic switch and circuit diagram
Courtesy of Allen-Bradley Co.

a start button and a stop button. When the start button is pressed, the magnetic switch operates to connect circuit wires to motor leads. When the stop button is pressed, the switch opens, disconnecting the motor. The means for accomplishing these results may be learned by referring to the schematic diagram, Fig. 25.

One end of the magnetic closing coil is attached to circuit wire 2. The other end is connected to one of the auxiliary contacts, and to a terminal of the start button. The wire from the other start button terminal connects to the stop button, and also to the second auxiliary contact. A wire from the remaining stop button terminal runs to a contact on the overload device. The second overload contact is joined with circuit wire *1*.

If the start button is pressed, current flows from wire 2, through the coil, across the start-button contacts to the stop button. It cannot flow down to the end of the magnet coil because the auxiliary contacts are still open-circuited. Current passes across the stop button contacts, therefore, and through the overload contacts to circuit wire *1*, completing the circuit and energizing the magnet coil.

As the switch closes, the auxiliary contacts meet, and current flows from wire 2 through the coil, the auxiliary contacts, the stop button, and the overload contacts, to wire *1*. When pressure on the start button is released, the contacts open without disrupting current flow through the magnet coil.

If the stop button is now pressed, the circuit through the coil is interrupted, and the switch opens. The same thing occurs if the overload contacts open, since this operation severs connection to wire *1*. Thus, the motor running protection is incorporated into the magnetic switch.

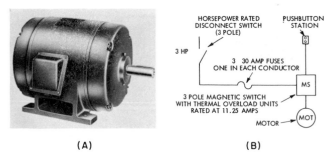

(A)   (B)

Fig. 26. Three-phase motor and diagram of three-phase motor installation
Courtesy of Wagner Electric Corp.

Connecting a 3-HP, 230-Volt, Three-Phase Motor

The motor in Fig. 26 is rated at 9 amps per phase. It is started on full line voltage. The circuit diagram is exactly the same as for the 1½-hp single-phase unit, except for the number of supply wires. A three-pole magnetic switch is used instead of a two-pole unit, but the connection to the push-button station is no different from that of the single-phase device.

All motors larger than 2 hp require horsepower-rated disconnecting switches. A 3-hp switch is needed here. The rating of the branch circuit protective device should not exceed 3 × 9 amps, or 27 amps. The nearest standard fuse is 30 amps, this size being acceptable. Motor running protection cannot be larger than 1.25 × 9 amps, or 11.25 amps. At least two thermal units of this value are required. NEC Table 310-12 lists the nearest size of Type R copper wire as No. 14 AWG.

REVIEW QUESTIONS

1. Is a range circuit two-wire or three-wire?
2. Can nonmetallic sheathed cable be used for a range circuit?
3. How is an electric range circuit terminated?
4. How is the frame of a clothes dryer grounded?
5. Does the home type of water heater usually have one, or two, heating elements?
6. A central electrical heating unit is similar to what other heating device?
7. How are branch duct heaters controlled in a duct-heating system?
8. Does the term *radiation* apply to air heated by circulating currents?
9. Are asbestos-insulated supply conductors needed for baseboard heating strips?
10. May heating cable be applied directly to a wooden surface?
11. Can heating cable be embedded in a concrete floor?
12. Are bathroom ceiling heaters usually installed on special circuits?
13. Do all thermostats operate on low voltage?
14. Would you mount a thermostat on the wall directly above a baseboard heating strip?
15. Must unit heaters be completely electric?
16. Does a heat pump make use of resistive elements?
17. Are heating calculations easy or difficult?
18. Enumerate the four essential calculations in motor wiring.
19. What is the maximum rating of a running-protective device for a 50-degree motor whose full-load current is 10 amps?
20. Is the stop button connected in the starting circuit of a push-button station?

Chapter ⑪

Multi-Family Dwellings,
Special Construction Features ▶

QUESTIONS THIS CHAPTER WILL ANSWER

1. What types of wiring are commonly acceptable for apartment house installation?

2. State how conduit is installed in a flat-deck slab?

3. In what respects does a pan-type installation differ from that of a flat-deck installation?

4. Outline the service problems likely to arise in connection with a three-story apartment building.

5. Where is stacking resorted to in an apartment house conduit installation?

Wiring Methods

Basically, the problem of wiring a multi-family dwelling is no different from that of a single-family unit. Because of the greater risks involved in multiple occupancy, however, wiring details are subject to greater restrictions than with the simpler structure.

Knob-and-tube installations are seldom used for apartments and motels, even where the inspection authorities allow it in single-family residences. Nonmetallic sheathed cable is frequently employed, especially in less populous areas. One large city permits nonmetallic sheathed cable in motels, and in apartment houses with four or less occupancies. Another city limits this material to one- and two-family dwellings. In neither city may this form of cable be installed in a hotel.

Armored cable is used rather widely in multi-family units. The use of a continuous bonding strip, which the Code now requires

174

in contact with the armor, has removed the most serious objection to this type of wiring. The strip insures positive grounding continuity when the cable is properly installed.

Electrical metallic tubing (EMT) may be used in all types of dwellings. Because of high labor costs, rigid conduit is seldom found today, except for underground runs where permanent moisture is encountered. Flexible metallic conduit is specified upon occasions. It is not so popular as a few years ago, when EMT was not yet introduced. Certain inspection departments limit application of this material to remodeling of existing systems.

Construction Procedures

Working procedures in frame buildings are similar to those for smaller units. Outlets are to be spotted, boxes set, holes drilled, cable or conduit run, and splices made. If there are a number of identical occupancies, prefabrication is often resorted to. This method, called, "prefab" by electricians, is based on the principle of mass production.

For example, suppose that the occupancy consists of twelve apartments with identical physical arrangements, and that armored cable wiring is specified. First, outlet boxes are spotted and holes drilled. Then, lengths of individual runs are measured between outlet and outlet, or outlet and distribution panel. Twelve pieces of each size are prepared.

Preparation involves cutting the numerous lengths, removing enough armor at either end to allow the proper amount of exposed conductors, reaming sharp ends of armor, and skinning the wires for splicing or for connecting to switches. If holes are drilled large enough to permit connectors to slip through, insulating bushings and connectors may be installed at the same time. By lettering or numbering the various pieces according to their respective locations, twelve separate bundles can be made up, one for each apartment.

When the bundle is delivered to a particular apartment, wiring may be installed quite rapidly. Time saved through prefabrication depends upon the number of occupancies concerned. Where there are four or less units, the method offers only slight advantage. It is most valuable when there are a large number of uniform living quarters, such as a housing project. The method can be employed also with nonmetallic sheathed cable, but the greatest possible savings are obtained where electrical metallic tubing or rigid conduit is used. This is particularly true of slab work, which is explained next.

Concrete Floors. Reinforced concrete floors come under the general heading of *slabs,* and the installing of conduit in such floors is known as *slab work.* Electricians also designate the concrete as *pour,* speaking of this type of work as a *pour job.* The Code allows either rigid conduit or EMT for this application. Certain inspection authorities, however, prohibit EMT here.

Ground Slab. The simplest type of slab is one poured directly on the earth. Figure 1 shows the floor of a motel room which has a plug receptacle at the middle of each wall, 6″ above the floor line. The circuit drops from a panel in the north wall down to a handy box. A piece of rigid conduit extends from this point to a 4″ square outlet box in the east wall, a right-angle bend being made at each location, and the conduit lying upon the earth. A second piece of conduit extends from here to another 4″ square box in the south wall. The third conduit runs from this outlet to a handy box in the west wall, the circuit terminating at this location. Wire screen is placed on top of the conduit by concrete workers.

The electrician first locates exact spots for centers of outlets, taking measurements from blueprints and transferring them to the ground by means of stakes or other markers. Since rigid galvanized conduit is employed, it may rest upon or in the earth. It is a good plan to install the outlet boxes, and to fasten them to wooden stakes, short pieces of reinforcing steel, pieces of conduit, or similar materials, as in Fig. 1, to insure that the conduit does not move during pouring operations. Conduit openings should

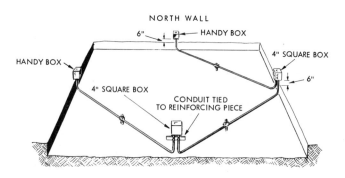

Fig. 1. Rigid conduit installation

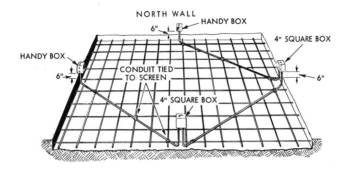

Fig. 2. EMT installation

be sealed by means of plugs or "pennies" so that concrete will not enter. It is then necessary to install single-receptacle plaster rings on the 4″ square boxes before the walls are erected.

When EMT is used, it should be surrounded on all sides by at least 2″ of concrete to guard against damage from permanent moisture. This requirement is met usually by installing the tubing after the wire screen is laid, as shown in Fig. 2. The tubing is placed on top of the screen and tied to it by soft iron wire. When the screen is raised from the ground by concrete workers, the tubing comes up also, and a bed of concrete flows underneath. After concrete is poured it is permitted to "set" before additional work is done.

In some cases, copper heating pipes are laid in the slab. Rigid conduit should be used in this case, installing it two to four inches below the heating pipes, depending upon the maximum operating temperature.

Upper Floor Slabs

Flat Deck Slab. There are two general methods of placing concrete in buildings which have one or more floors above ground level. The flat-deck scheme makes use of a smooth surface, or deck, of plywood upon which reinforcing steel is spread before concrete is poured. The slab may be four, six, or eight inches thick, depending upon structural design, and the reinforcing steel is "woven" in two or more layers.

The electrician marks outlet locations on the smooth deck by means of a crayon or chalk. Ceiling outlets of rooms underneath are included, Fig. 3, if the concrete slab is to act as the finished

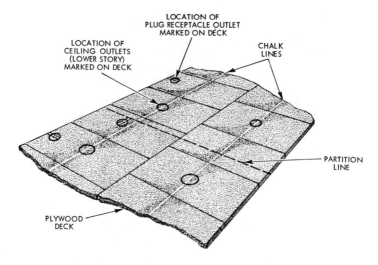

Fig. 3. Location of outlets marked on flat deck

ceiling of the lower story. Partition lines, too, are sometimes
marked. Concrete boxes, like the one shown in Fig. 4, are employed
for the lower floor ceiling outlets. This box consists of an octagonal
ring and a cover plate. The box has projecting ears for nailing it to
the deck, and the plate may be equipped with a fixture stud
if desired.

Rings are fastened to the deck, if possible, before the first layer
of reinforcing steel is spread. After the steel is in place, rigid
conduit or EMT is run from box to box. The term *conduit* will
include both rigid conduit and EMT throughout the remainder
of this chapter, unless one or the other type is specifically mentioned.
Conduit for plug receptacles is extended from panel locations to

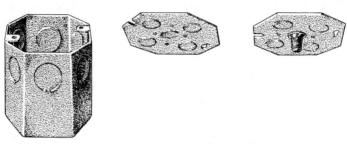

Fig. 4. Concrete boxes and cover plates

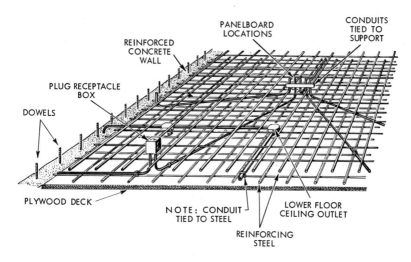

Fig. 5. Flat-deck conduit installation

the various outlets in much the same way as in the ground slab. Plug receptacle boxes are often supported by metal brackets which are nailed to the deck.

In Fig. 5, conduits extend from the panel location to plug receptacle boxes. A feeder conduit is also shown. Conduits are grouped to enter the bottom of a distribution panel, which is to be installed later. The group is tied to a piece of wood or other support to hold it firmly in place.

Note the run which extends from concrete box A toward the wall. It is coupled to a right-angle bend formed in a length of conduit that was placed in the wall before the decking was laid. This conduit drops to a switch outlet or a panel location on the floor below, so that ceiling outlets may be controlled from there.

When this portion of the work has been completed, steelworkers install the next layer of reinforcing steel. At this time, the electrician prepares the boxes and ties the conduits to the steel with iron wire. This procedure is necessary to prevent undue movement or damage to the conduit in the pouring operation. The matter of preparing the boxes consists of taking steps to make sure that concrete will not harden inside, thereby rendering them useless.

If the builder permits, ¾″ holes are bored in the decking at the center of each box. Any wet material that enters can then drain out before it hardens. In case this method is prohibited, boxes may be filled with paper, sand, or other suitable material before cover plates are installed. The filling material is readily cleaned out after the decking is stripped from the concrete slab. Where sand is used, ends of conduit projecting into the boxes must be sealed off. It is usually advisable to do so, regardless of the method of preparing boxes.

Considerable saving in time is possible, where local inspection authorities allow EMT in concrete, especially if it is used with crimp or tap-on fittings. But, when other craftsmen are roughshod in treatment of EMT stubs which project above the surface after the concrete has hardened, much of the time saved will be squandered in repairing flattened or broken ends.

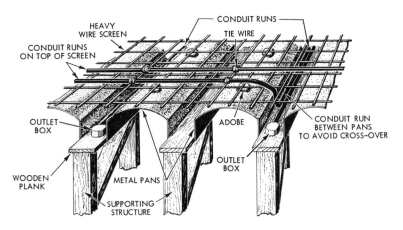

Fig. 6. Pan-type deck showing conduit runs

Pan Slabs. The pan type of construction is illustrated in Fig. 6. As shown, horizontal planking is laid upon supporting structures at regularly spaced intervals. The metal pans used here are attached to the edges of planks by nailing, and are overlapped at the ends so as to provide a complete form or mold that will retain wet concrete.

After pans are set, the electrician installs outlet boxes on the wooden planks to support fixtures which light the story below. Conduit or tubing is run from box to box, box to switch, or box to panelboard. Conduits that lie upon the planks are raised an inch or so by patented supports or by pieces of *adobe*. This material, used by the steelworkers, comes in narrow strips or blocks which are made of light-weight concrete.

Workmen then place reinforcing bars in the space between the pans, and lay comparatively light bars or screen on the pans to form a deck. The electrician follows, running conduit on top of the steel. Since the floor slab in pan construction is relatively thin, it is usually bad practice to have conduits cross one another. To avoid doing so, cross runs are made in the space between pans, as shown in Fig. 6. Conduits are tied to reinforcing bars or screen with iron wire. Outlet boxes are prepared in the same way as in flat-deck construction so they will not fill up with concrete.

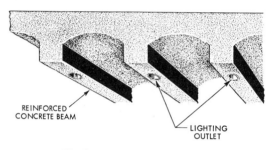

REINFORCED
CONCRETE BEAM

LIGHTING
OUTLET

Fig. 7. View of pan-type slab

Figure 7 illustrates a section of completed floor that has been stripped of metal forms and their wooden supports. Concrete in the spaces between pans has formed reinforced beams, and a thin floor web stretches between them with lighting outlets extending along the face of a beam.

Both types of slab are in common use. In some buildings, the first floor is pan-constructed, while upper ones are of the flat-deck type. Insofar as the electrician is concerned, the flat deck offers lesser problems.

Concrete Walls. Electrical outlets in reinforced concrete walls or partitions are handled the same way as those in floor slabs. In Fig. 8A, a plug receptacle, a lighting outlet, and a switch outlet are to be installed in the wall, the receptacle facing inward, the light outward, and the switch inward. The outer half of the wooden forms has been set in place, and the reinforcing iron fabricated. The conduit has been tied to reinforcing iron exactly as in floor slabs.

Before the inner half of the wooden forms is set in place, boxes

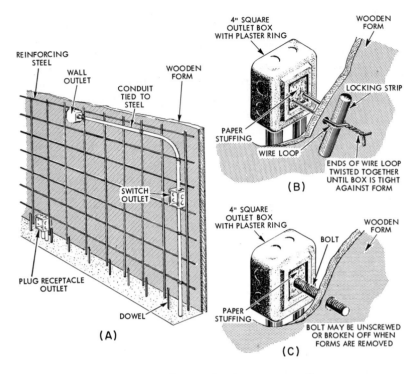

Fig. 8. Installing outlets in concrete wall

are filled with paper, and an open loop of iron wire is inserted through screw holes in the back of each box, Fig. 8B. The purpose of the loop is to hold the box tightly against the form, where it belongs, and to exclude wet concrete. A small hole is bored in the wood so that the iron wire loop may come through. A locking strip of reinforcing iron, conduit, or even wood, is placed between the open ends, on the outer surface of the form, before they are twisted together.

An alternative but more expensive method, Fig. 8C, is to use a center bolt instead of a wire loop for securing the outlet box to the wooden form.

Chases. One additional measure connected with this type of operation is that of arranging for openings in the floor to accommodate vertical conduit risers. Figure 9 shows how it is done. If a rectangular hole is desired, so as to allow several large conduits to be placed side by side, a wooden box is constructed as in Fig. 9A.

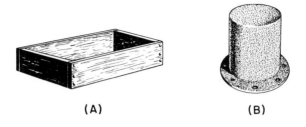

(A)   (B)

Fig. 9. Box and sleeve chases

The box is nailed to the wooden deck and filled with sand. It is easily removed, after the concrete floor has set, leaving an opening or chase. If a hole for a single conduit is desired, a metal sleeve or "can" is fastened to the deck, as in Fig. 9B. Such openings are often made in closets or other out-of-the-way places. After the conduits are installed, openings around them may be filled in with cement or "grout."

Service Considerations

Service requirements for multi-family units differ in some respects from those of single-family residences. They usually employ larger conductors, the location of service disconnecting means is important, and the service entrance conductors on large jobs are usually brought in underground. Methods of connecting underground service conductors to the power company mains have been considered earlier.

Single-family dwellings are always supplied by either two-wire or three-wire single-phase current. Apartment houses and hotels are often supplied by three-phase current, usually in the form of a network. As mentioned earlier, the network is a four-wire arrangement that has three phase wires and a neutral conductor. The voltage between any two of the three phase wires is 208 volts, while that between any phase wire and the neutral conductor is 120 volts. Lighting panels in the various apartments are usually supplied by a three-wire feeder consisting of two phase wires and the neutral wire. Load on the main service conductors is equalized by making sure that the same number of panelboard taps is taken from each of the three wires.

The Code provides that each tenant of a multiple-occupancy building shall have access to his own disconnecting means. It also

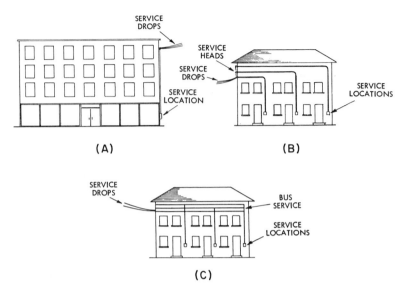

Fig. 10. Multi-unit services

states that a multiple-unit building having individual occupancies above the second floor shall have service equipment grouped in a common accessible area. Figure 10A shows a multi-story apartment building. To comply with the Code, individual service switches must be grouped at a central point. Here, the selected location is on the outside, at one end of the building. It could have been placed in a more protected spot, if desired, such as in the entrance lobby or in an unlocked meter room.

When there are no apartments above the second floor, Fig. 10B, separate service entrance conductors may be tapped from one set of drops. Here, three separate conduits are run to the spot where service drops are attached to the building. The arrangement of Fig. 10C is also permitted. In this case, the drops furnish current to service wires that extend across the building. Sub-service conductors for individual locations are tapped from these wires. This latter arrangement is sometimes known as a "bus" service.

The presence of fire escapes on exterior walls of hotels and apartment houses requires that special attention be given to the location of service drop conductors and the service head. Figure 11 illustrates a situation often met with in practice. The service drops may not be closer than 3' to a window unless above its top

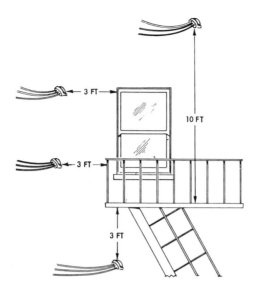

Fig. 11. Minimum clearances of service drops

level. In such case, the drops are considered out of reach from the window. Drops must clear the fire escape platform not less than 3′ horizontally, 3′ vertically below, or 10′ vertically above.

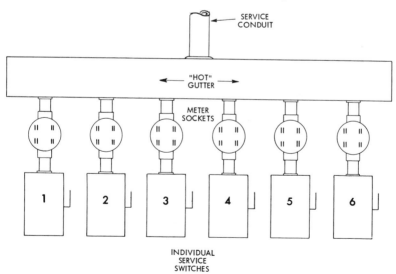

Fig. 12. Built-up switchboard

Switchboards and Panelboards

The Code allows not more than six switches or circuit-breakers to act as the service disconnecting means. Figure 12 shows a built-up arrangement for six meters and service switches. Entrance conductors run from the service head to the "hot" gutter. Individual services are tapped to main conductors in the gutter.

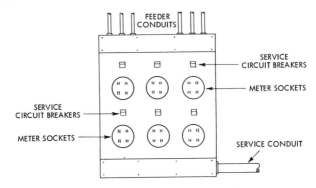

Fig. 13. Commercial switchboard

A commercial switchboard that accomplishes the same purpose is pictured in Fig. 13. Meter sockets and service switches are mounted in a rectangular enclosure. The service conduit enters at the lower right side, while individual feeder conduits extend from the upper surface of the metal box.

With more than six individual services, a main disconnect switch is required. Figure 14 shows a switchboard that has eight meter sockets, eight disconnect switches, and eight feeder conduits. This

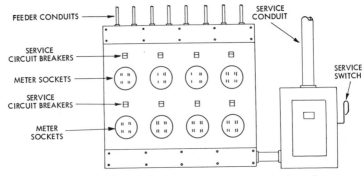

Fig. 14. Switchboard for more than six individual services

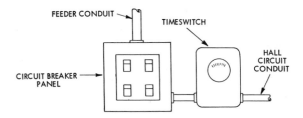

FEEDER CONDUIT

TIMESWITCH

HALL CIRCUIT CONDUIT

CIRCUIT BREAKER PANEL

Fig. 15. "House" panelboard with time switch

arrangement could be used for eight apartments or for seven apartments and a house feeder. The house disconnect switch may control a panel that has branch circuits for hall lights, a washing machine, and perhaps a dryer. Figure 15 shows a house panel with circuit breakers, and also a timeswitch that turns hall lights on and off at preset intervals.

The timeswitch has a mechanism which operates a pair of switch contacts through a train of gears. A clock motor drives the gear train. An adjustable arm on the face of the dial may be set for the hour at which it is desired to close the circuit, and another arm for the hour at which the circuit is to be opened. A common practice is to set the ON arm for 6 P.M. throughout the greater part of the year, and the OFF arm for 6 A.M.

Whether or not apartments and motels are separately metered, there is usually a circuit panel inside each occupancy. Hotels often have large panels located in closets on each floor, the panel supplying a number of rooms or suites in its immediate vicinity.

Stacking

Telephone or television outlets in apartment houses and hotels are usually stacked as in Fig. 16A, the conduit from the outlet in one apartment continuing upward to an outlet in another apartment directly overhead. It may be noted that the telephone stack on the right extends upward from the basement, while the television stack on the left descends from the roof. There will be as many separate stacks, in general, as there are apartments on a single typical floor. Telephone stacks will be run to junction boxes in the basement, and television stacks to junction boxes in the attic or on the roof.

Lighting panelboards are sometimes stacked as in Fig. 16B. Those on the first and second floors have wiring gutters so that leads from buses may be attached to the feeder conductors. This plan is used mainly where electricity supplied to the apartments is not separately metered. The general scheme may be employed, even where separate meters are installed at the service location, if individual feeder conductors are run to each panelboard.

Conduit entering the lower panelboard in this case will have nine wires, assuming that three-wire panels are used. Six conductors will extend to the second floor panelboard, and three will continue to the one on the upper floor. It should be noted that no wiring gutter is needed for the upper panelboard. There are, of course, as many stacks as there are apartments on a single floor.

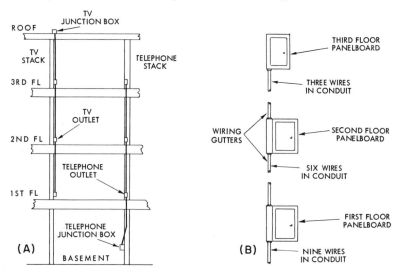

Fig. 16. Stacking

Emergency Lighting

In many inspection localities, hall lighting circuits in multi-story apartment houses and hotels are classed as semi-emergency circuits. As such, they are required to be installed in conduits entirely separate from other wiring, such as that for utility plug receptacles in common hallways. In some instances, switches that control these circuits must be locked. These are not really emergency circuits within the Code definition, however, unless they are arranged for automatic power supply from a special emergency service in case of necessity.

REVIEW QUESTIONS

1. Are knob-and-tube installations commonly found in apartment houses and hotels?
2. Is armored cable objectionable in this application?
3. Upon what principle is prefabrication based?
4. Would prefabrication result in considerable labor saving if used in a three-apartment building?
5. Does the Code permit EMT in slab work?
6. Should EMT be used, generally, where copper heating pipes are cast in the slab?
7. Of what two parts is a concrete box made?
8. How should a group of conduit stubs be treated at a panelboard location?
9. How is undue motion of conduit runs prevented during pouring operations?
10. Why are concrete boxes sometimes filled with sand?
11. Is a hole usually drilled in the plywood deck before filling the box with paper?
12. Where are ceiling outlet boxes generally located in the pan-type slab?
13. Is heavy reinforcing iron used in the floor slab of the ordinary pan-type job?
14. How are plug recepatcle boxes fastened to wall forms?
15. What term applies to holes cast in a poured floor?
16. May service conductors in a three-story apartment building be run directly to each occupancy?
17. What clearance from service drop conductors shall be maintained above the deck of a fire escape?
18. Does the Code require a main disconnect switch for six individual services?
19. How are TV outlets usually arranged?
20. Are hall lights of apartment houses always installed on Code-standard emergency circuits?

Chapter (12)

Residential

Furnace Controls

QUESTIONS THIS CHAPTER WILL ANSWER

1. What electrical controls are required for a hot-water heating plant?
2. What electrical control devices are used with an air-heating system?
3. What is meant by stoker controls?
4. Are any special precautions necessary in wiring an oil-burner system?
5. What control units are commonly associated with a gas-operated heating plant?

Automatic Control Equipment

Besides installing electrical heating devices, the electrician is called upon to wire for electrical accessories to coal-, gas-, or oil-heating systems. The most common automatic control unit applicable to all types of heating is the thermostat which has been discussed in the preceding chapter. There are a number of other electrical control devices, but most of them apply to only one system.

Aquastat. Used in connection with hot-water heating plants, the aquastat is a temperature-sensitive switch that governs a control circuit to shut off the heating system as the water in the boiler reaches a set temperature. There are two general types of these controls; namely, the immersion type, Fig. 1 and the surface type, Fig. 2. The *immersion* unit has a tube which screws into the top or side of the boiler, or into a fitting in the hot-water riser located as near the boiler as possible.

The *surface* or *clamp-on type aquastat* is strapped to the hot-water riser and depends upon transfer of heat from the pipe to its operating element. It is used when it is impossible to drill a hole in

Fig. 1. Immersion type of Aquastat   Fig. 2. Clamp-on Aquastat

Courtesy of Minneapolis-Honeywell Regulator Co.

the boiler or riser. Because of air currents and resultant temperature variations around the boiler this device is not as satisfactory as the immersion type aquastat.

Pressurestat. The pressurestat or pressuretrol, Fig. 3, performs the same duty with respect to the steam boiler as the aquastat does with respect to the hot-water boiler. It opens the control circuit to shut off the heating equipment, when the desired pressure has been reached. This pressure, in a residential installation, is usually about five pounds. The device must be screwed into the boiler so that its temperature-actuating bulb is in direct contact with the steam.

Airstat. This unit provides the high-limit protection essential to the operation of a warm-air furnace. Bolted to the dome of the furnace, with the operating tube inside the bonnet, the airstat opens the control circuit when air temperature inside the furnace reaches a predetermined value.

Furnacestat. This device is used with a warm-air furnace that has a circulating fan in the duct system. Figure 4 shows a furnace-stat used to prevent circulation of cold air through the rooms. It maintains an open fan circuit until the temperature of the air in the bonnet reaches a preset minimum value. Since this device, like the

Fig. 3. Pressuretrol   Fig. 4. Furnacestat
Courtesy of Minneapolis-Honeywell Regulator Co.

airstat, is located in the furnace bonnet, it is possible to obtain a combined unit known as a *combination furnace controller,* Fig. 5.

Motor-Operated Draft Control

Now that most of the important heat-control equipment has

Fig. 5. Combination furnace control
Courtesy of Minneapolis-Honeywell Regulator Co.

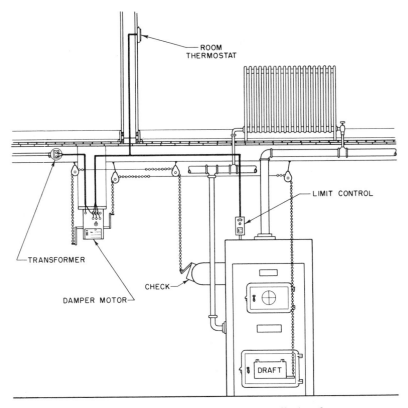

Fig. 6. Illustrating a damper-control installation for a
hot-water or steam boiler

been briefly explained, the installation of draft control on hand-
fired boilers will be considered. A draft-control system includes a
motor which rotates a half-turn in either direction to open or close
the dampers according to temperature demands. Figure 6 shows a
damper-control installation for a hot-water or steam boiler; and
Fig. 7 shows the same type of installation on a warm-air furnace.
Both connections in Fig. 6 and 7 are alike except for differences in
the limit controls.

On some installations, a 110-volt motor operates the dampers,
while a 24-volt transformer supplies the control circuit.

Care should be taken to locate the motor in a dry, clean place,
and in such a position that the chains between motor arms and
dampers move freely on the pulleys. The connection diagram for

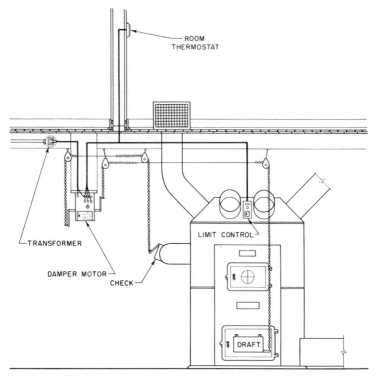

Fig. 7. Illustrating a damper-control installation for a
warm-air furnace

a damper-control low-voltage installation is shown in Fig. 8. The
letters *R, B,* and *W* refer to colors of insulation on the wires—red,
black and white.

For automatic heat-control wiring connected to the 120-volt

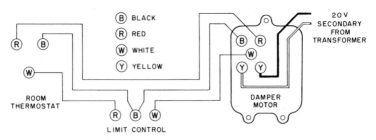

Fig. 8. Diagram of connections for a damper-control low-
voltage installation

service, No. 14 wire is adequate; for 24-volt control wiring, No. 18 thermostat wire should be used.

INSTALLING HEATING-SYSTEM CONTROLS

Stoker Controls

In the foregoing discussion of damper controls, the only concern was to open and close the dampers in accordance with demand for heat. The fire had to be maintained clinker-free if the damper controls were to function properly. With a stoker, furnace coal feed and forced draft are two additional automatic features which must be handled by the controls. Figure 9 shows the equipment layout and wiring diagram for a stoker installation. This scheme can be used for warm air, hot water, or steam heat, if the correct form of high-limit control is chosen.

Coal will not burn long in the absence of oxygen. Since the only supply of air is by way of the forced draft unit, a stoker relay, or timer, is employed. This device causes the fan to send a blast of air through the combustion chamber at spaced intervals. Such action enables the fire to continue burning during comparatively warm periods when the house thermostat may not call for heat.

From the diagram shown in Fig. 9, it is evident that line voltage is used for the high-limit control, the stoker relay, and the stoker motor. Low-voltage wiring runs from the thermostat to the stoker relay. The dotted lines indicate a low-limit control switch. This relay is used to maintain boiler water at a temperature sufficient for domestic use in the nonheating season.

Oil-Burner Controls

The installation of oil-burner controls is complicated by the addition of a stack-relay switch. This switch or *protector relay,* as it is sometimes called, is installed in the flue pipe. It functions to shut off the burner in case the stack does not come up to a certain predetermined temperature within 45 seconds after the motor starts. The relay prevents flooding of the basement with oil in case the ignition system fails.

Ignition of the oil is accomplished with the aid of a spark gap which is connected to the high-voltage terminals of a transformer installed on or near the burner.

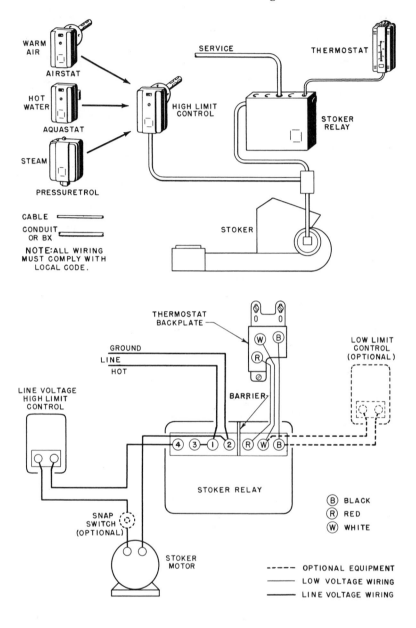

Fig. 9. Diagram of connections for a stoker installation

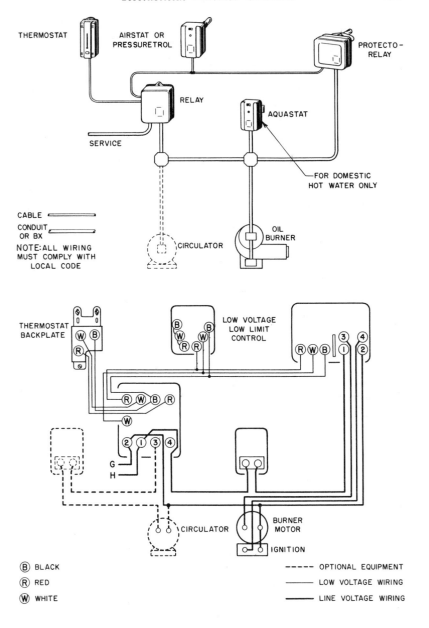

Fig. 10. Diagram of connections for an oil-burner installation

The spark may be continuous, or it can be made to cease when the oil starts to burn. Certain factors will determine the correct method. The continuous spark system is generally advised when down drafts are common in the particular location. With a continuous spark, it is necessary to replace the electrodes more often than with the intermittent spark.

After the spark has been shut off, the oil continues to burn because the temperature of the firebrick lining is high enough to ignite the oil.

The conduit layout and diagram of connections for an oil-burner installation is shown in Fig. 10. Note that supply wires are connected directly to the motor-starting relay housing which also contains the low-voltage transformer. The low-temperature control and circulator, indicated by dotted lines, should be used on a forced-feed hot-water system only, and not with a steam or hot-water system.

Gas-Burner Controls

The main factor to bear in mind when installing automatic controls for gas burners is that a dangerous quantity of gas must not be permitted to accumulate in the combustion chamber. There is a variety of safety devices for dealing with this hazard. In general, there are two means of feeding gas to the boiler when the room thermostat calls for heat: by means of a diaphragm valve, or by means of a solenoid-operated valve. The diaphragm valve is pressure actuated. A solenoid coil opens a small valve which permits gas to escape from the top of the diaphragm. When this occurs, normal pressure underneath raises the diaphragm, and gas flows into the

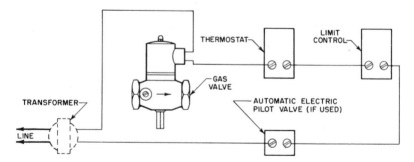

Fig. 11. Gas-burner controls

boiler. When a solenoid type main valve is used as in Fig. 11, completion of the operating circuit energizes the solenoid coil and pulls up a plunger which in turn opens the valve.

One difference between gas-burner and stoker or oil-burner controls is that all the gas-burner devices operate on low voltage. This is possible because no motor-driven fuel pump is required for a gas unit.

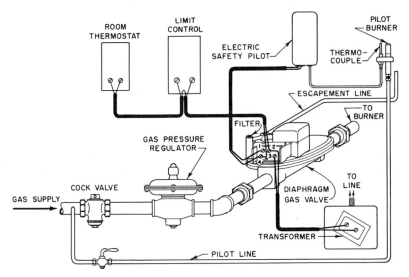

Fig. 12. Illustrating a gas-burner installation

Figure 12 shows a layout of the relative positions of equipment needed for a gas-burner installation. Note that the electric safety pilot is wired to a thermocouple. If the pilot burner is extinguished, no current will be generated in the thermocouple element, thus causing the circuit for the diaphragm- or solenoid-operated valve to remain open.

All protective devices should be so designed as to terminate operation of a unit without use of electricity. Hence in the event of current failure the fuel supply is automatically cut off.

REVIEW QUESTIONS

1. For what is an aquastat used?
2. What are the two types of aquastat?
3. Which type is more satisfactory?

4. What is a pressurestat?

5. What is the usual steam pressure in a residential installation?

6. How is the pressuretrol fastened to the steam boiler?

7. What is an airstat?

8. At what point on the equipment is it installed?

9. What is a furnacestat?

10. Where is it placed on the equipment?

11. What is a combination furnace controller?

12. How are automatic draft dampers usually operated?

13. What voltage is used customarily for control units of an automatic draft-damper installation?

14. What is the smallest size of conductor permitted for wiring 120-volt valves?

15. What is the smallest size of conductor permitted for wiring 24-volt controls?

16. What is a stoker relay?

17. State another common name for the stoker relay.

18. What is the specific purpose of an oil-burner protector relay?

19. What two types of ignition are employed in connection with oil burners?

20. What are the two general methods of feeding gas to a boiler when heat is called for by the thermostat?

Remote Control

Wiring

QUESTIONS THIS CHAPTER WILL ANSWER

1. *What is the principle underlying remote control wiring?*
2. *What are the advantages of remote-control wiring?*
3. *How does the remote-control relay operate?*
4. *Is gang-mounting of relays superior to zone-mounting?*
5. *Is it possible to turn on all house lights from one point?*

PRINCIPLES OF REMOTE-CONTROL WIRING

How Remote-Control Units Operate

The purpose of remote control in an electrical circuit is to govern the operation of a current-consuming device at some point removed from the device. A wall switch, in one sense, is a form of remote control for a ceiling outlet, because the light is turned on and off at a point that is some distance away.

The remote-control devices to be considered here are different from the previous example in that low voltage is used. To show how this is accomplished, examine Fig. 1. This is an ordinary snap switch, like the wall switch of the original example. The switch knob, however, is made of soft iron. If a coil is placed above the switch knob, the knob will be drawn upward when the coil is energized. Here, a battery is used as the source of supply for the remote control unit. The coil is energized by closing the push button, which can be placed at a distance from the switch. When the coil is energized, the snap switch moves to the ON position, and remains there, so that it is no longer necessary to energize the coil.

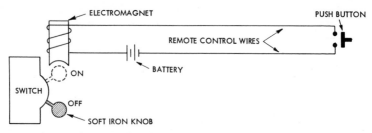

Fig. 1. Schematic drawing, showing principle of low-voltage
remote control

Another circuit can be used to turn the switch off. These two
circuits are shown in Fig. 2A. When the upper push button is
pressed, the switch is turned on; when the lower push button is
pressed, the switch is turned off. This circuit can be greatly simpli-

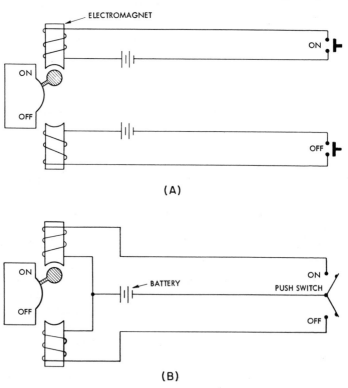

Fig. 2. Wiring method for switching ON and OFF by remote
control

fied by eliminating one battery and combining the two push buttons into one unit, as in Fig. 2B. Inspection of this circuit will show that, when the ON button is pressed, current flows through the upper coil, turning the switch on. The switch is turned off when the circuit through the lower coil is energized.

Use of Remote-Control Units in Wiring Systems

The first step in applying remote control to a system is to convert it to alternating-current operation; otherwise, necessity for battery replacement would be undesirable. A step-down transformer is used to reduce the voltage from 120 volts to 24 volts. There is an additional refinement in the alternating-current unit, namely, the exciting coils are combined into one unit with a center tap. Simplifying the snap switch results in a reduction in size.

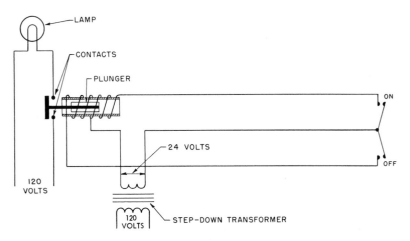

Fig. 3. Use of alternating current for remote-control switching

The basic circuit for alternating-current operation is illustrated in Fig. 3. By connecting one side of the transformer to the center tap of the coil, either half of the coil can be excited at will. For example, pressing the ON push button energizes the right-hand half of the coil, moving the plunger to the right. This closes the contacts in the lamp circuit, causing the lamp to light.

This simple switching arrangement can be extended so that an outlet can be controlled from a number of different points, by placing the desired number of push buttons in parallel. Circuit

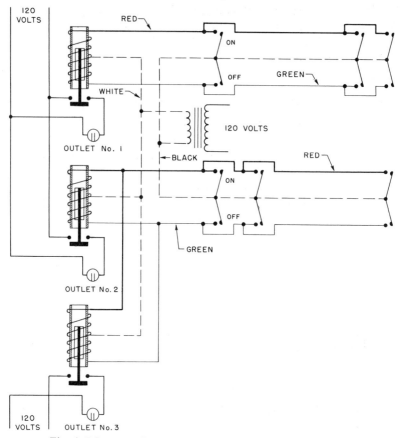

Fig. 4. Diagram of remote-control circuit for control from
various points

arrangements for two sets of push buttons are shown in Fig. 4.
The first outlet is controlled by three push buttons in parallel. This
outlet may be a single light or a whole circuit. The two outlets
indicated in the lower portion of the illustration can also be
controlled by several push buttons. The fact that they need not be
on the same branch circuit greatly increases the flexibility of this
wiring system.

Advantages of Remote-Control Wiring

Some of the advantages of this system are already evident,
but there are still others worth noting. A major consideration in

any wiring system is safety. The low-voltage system results in safer operation by removing high voltages from control points. A further advantage is the ease of switch relocation. In addition, a selector switch can be arranged to turn off all lights from a single location. In this way there is no likelihood of having lights on all night in some remote part of the home because they were overlooked.

EQUIPMENT FOR REMOTE-CONTROL WIRING

Transformer

The transformer used in remote-control wiring, Fig. 5, has a 120-volt primary winding and a 24-volt secondary winding. The mounting plate is constructed so that it can be attached to a standard square outlet box. The primary, or 120-volt, leads come out through the mounting cover, which permits them to be connected directly to the supply within the box. The secondary terminals can be seen on top of the transformer case. These transformers have an additional advantage that is worth noting. This is the current-limiting feature, which protects the windings from overload. Therefore, no overcurrent protection is needed in the secondary circuit; even a direct short-circuit will not cause the transformer to overheat.

These transformers are built to operate several relays at one

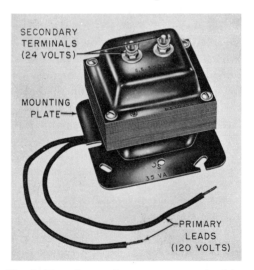

Fig. 5. Transformer for remote control wiring
Courtesy of General Electric Co.

time, the number depending upon the size of the particular unit.

The run from the push button to the relays must now be considered. The capacity of the transformer and the current drain of the relays must be taken into account when arranging a wiring plan. Manufacturers of remote-control equipment offer this information. As an example, three or four relays may be operated from a distance of about 100′. Accordingly, a lesser number of relays can be operated from a greater distance until the limiting distance for operating a single relay is reached. Generally, this is in the vicinity of 1000′ to 2000′, depending upon the individual transformer, relay, and conductors. This distance, however, is usually great enough to handle any switching problem in the home or on the farm.

Relay

The relay used in remote-control wiring is a single-pole switch that is set in the ON or OFF position by exciting the proper coil element. Once the relay is set in either position, it remains there until the other coil element is energized. For this reason it is unnecessary to maintain excitation after switching is accomplished. Therefore, the switches that control the exciting current need be only momentary-contact units, or push buttons. The 24-volt leads and the 120-volt leads come from opposite ends of the unit. The barrel of the relay is inserted through a ½″ knockout from the inside of the outlet box.

This arrangement isolates the low-voltage wiring from the high-voltage wiring, as required by the National Electrical Code. Switching in the high-voltage circuit is handled within the box, while low-voltage control equipment is on the opposite side of the separator formed by the outlet box. This construction allows the relay to be placed at the desired switching point without additional mounting and isolating devices.

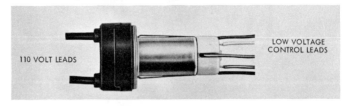

Fig. 6. Remote-control relay
Courtesy of General Electric Co.

Figure 6 shows the construction of one particular relay. The high-voltage leads and their associated contacts are built into the portion of the relay remaining within the outlet box. The solenoid and low-voltage leads are set in the opposite end.

Control Switches

The control switches, which are constructed for flush or surface mounting, are actually two push buttons built into one unit.

A plaster ring box cover is used to support flush-mounted switches, but the surface-mounted units can be installed almost anywhere. This is especially advantageous when switches are to be placed on thin partitions, brick-filled walls, or other mounting locations having limited depth.

Selector Switch

The selector switch provides one of the outstanding features of low-voltage, remote-control wiring. It permits control of any or all relays from a centrally located position; for example, the bedroom. This selector switch, Fig. 7, has the usual push button, and, in addition, it has a selecting device. The selecting device inserts the push button in circuit with any one of a number of different relays or groups of relays.

Fig. 7. Nine-point selector switch
Courtesy of General Electric Co.

The selector switch allows all circuits to be turned on or off from this point. Depressing the push button in the OFF position and rotating the selector through the several positions accomplishes this result. With proper wiring arrangements, it is possible to turn particular circuits off and others on with a single motion. This is done by reversing leads of units that are to be turned on when the OFF button is pressed.

Fig. 8. Three-conductor cable, showing identifying rib on upper conductor
Courtesy of General Electric Co.

Wire

Remote-control wiring can be done with No. 18 RF (rubber covered) or No. 18 TF (thermoplastic covered) wire. Since no additional protection is necessary, these conductors are run in the open through partitions, floors, and ceilings.

The push buttons have a marked ON-and-OFF position so that the proper section of the relay solenoid is excited to open or close the high-voltage contacts. For this reason, there must be consistency in the wiring procedure. One recommendation is to employ color coding. An example of this color code is noted in Fig. 4. A second method codes the wire, itself. A ribbed section on the outside of one of the wires of the group, Fig. 8, serves to identify all three conductors. Thus, the ribbed wire may be taken as the common one, the middle wire the closing circuit, and the lower wire for the opening circuit.

PLANNING REMOTE-CONTROL WIRING

Remote-Control Layout

As in any wiring project, the desired operation of electrical facilities should be determined in advance. The flexibility of remote-control wiring offers a wide choice of controls.

Switches placed at the entrance of the home can control selected lights inside, so that entrance is made to a well-lighted building. The front entrance can be lighted from additional points in the building; for example, the kitchen. Radio or television can be turned off from the telephone location. Reassurance is immediately gained in moments of necessity by illuminating home and yard from a switch in the bedroom. Exhaust fans as well as heating units can also be controlled from one or more remote locations.

Locating the Relay

There are three general methods of arranging remote-control

relays. Each has distinct advantages in the type of service to be rendered.

Outlet-Mounted Relays. This is probably the simplest method from the standpoint of installation. A saving in high-voltage wiring is also effected by placing the relay at the point of control. With this method, however, the individual relays are not readily accessible if servicing should be necessary.

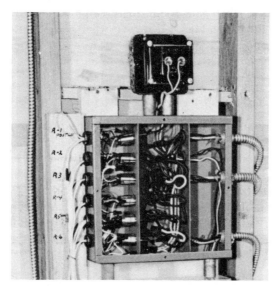

Fig. 9. Gang-mounted relays installed in attic

Courtesy of General Electric Co.

Gang-Mounted Relays. Here, all of the relays are mounted in one conveniently located panel. This scheme provides the simplest arrangement for servicing, which is an important consideration when a large number of units are to be operated. There is an additional feature to gang mounting, namely, it provides completely silent operation. With the panel box in an isolated location, it is impossible to hear the tripping of the relays. This mounting is shown in Fig. 9.

Zone-Grouped Relays. Arranging the relays in several groups affords some advantages. The residence, or other installation, is divided into areas in which the control of outlets is limited to four relays. This reduces the high-voltage runs as compared with the system employing a single panel box, yet retains most of the accessibility gained with the single panel. The usual procedure is to mount the relays in a centrally located housing, the size of which is determined by the dimensions of the particular relays. The box, in this case, provides the necessary shielding between high-and-low voltage sections.

Locating Transformer

Location of the transformer is not the same for the three relay arrangements. When gang mounting is employed, the transformer is situated at the panel, Fig. 9. However, when either outlet mounting or area grouping is employed, the transformer should be more centrally located. This is usually at some convenient and readily accessible point in the basement or attic.

Locating Selector Switch

Probably one of the most convenient locations for the selector switch is in the bedroom, since all lights may be properly set for the night before retiring. Variations in wiring the selector switch, as already noted, will turn on lights that should burn all night—gangway porch, hall, or bathroom—and turn off all other lights. A second desirable location for a selector switch is the kitchen, permitting control to be extended to other portions of the building without interrupting common household duties. The family room

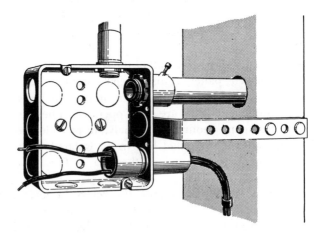

Fig. 10. Method of mounting outlet boxes with remote-control relays

is a third location that may warrant a selector switch, because it is usually the center of activity.

WIRING PROCEDURES

Installing Outlet Boxes

Outlet boxes are mounted in the conventional manner, since the use of low-voltage remote control does not affect supply wiring procedures. It does, however, simplify the switching of circuits. Figure 10 illustrates one acceptable method for mounting an outlet box. The box is held in place by a mounting strap attached to the joists. This illustration also shows two sections of conduit entering the box, and the remote-control relay in place.

Installing Switches

Switch installation in the low-voltage, remote-control system is greatly simplified as compared to the conventional wiring system. The switch can be mounted on a plaster ring box cover attached to the studding, Fig. 11. This ring will accommodate up to three low-voltage switches.

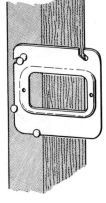

Fig. 11. Mounting-box cover

Installing Relays

Relays can be gang-mounted on a metal strip, Fig. 9, which serves also to isolate high-and low-voltage circuits from one another.

The method of installing outlet-mounted relays is shown in Fig. 10. The relays, however, should not be attached until the plastering has been completed. That is, all wiring is finished beforehand, except installation of relays and switches. These items are withheld to prevent damage during the plastering operation.

Installing Wire

Color or number coding can be used, or two- and three-wire cable, which has an identifying feature, may be employed in remote-control wiring. These systems vary somewhat. In one instance, black and white are used for the transformer secondary conductors. The white conductor leads to the common point on the relays (X, or No. 1), and the black conductor leads to the common point on the push buttons (X, or No. 1). Green and red conductors are used between push buttons and relays. The red connects to the ON terminal of the relay (C, or No. 2) and the ON terminal of the push button (C, or No. 2). The green conductor connects the OFF terminals (O, or No. 3) of the relay and switch.

With proper notation, the identified two- and three-wire cable can be used in the same way. Three-wire cable will be found

Fig. 12. Running multi-conductor cable through joists

Fig. 13. Stapling cable to studding with staple gun
Courtesy of General Electric Co.

especially useful in connecting push buttons in parallel and in making runs to the selector switch. Two-wire cable is most often used for runs from the transformer. These conductors must be run through joists and studs, Fig. 12. Wires used for remote control must be supported at 4½′ intervals usually by means of staples, Fig. 13.

Rewiring

Each remodeling job offers individual problems, and no single plan will satisfy all conditions. However, close examination of re- mote-control wiring will show many ways it can be used to advantage.

Surface-mounted switches make the low-voltage system advantageous for rewiring, because they can be mounted so easily. Furthermore, the small two- and three-wire cable can be run behind moldings or along baseboards, and through areas that would not accept armored cable.

If necessary to run cable across a plastered wall, it can be laid in a shallow groove and plastered over. A method of inserting an additional outlet box to handle the relay is shown in Fig. 14. This procedure is used where the original box will not hold the relay,

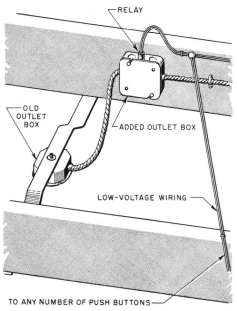

Fig. 14. Use of additional outlet box for mounting relay

which is frequently the case with older wiring equipment. When it is desirable to replace a single wall switch with several distributed push buttons, the low-voltage leads from the new push-button locations are connected to the operating coil of a relay. The 120-volt leads that went to the old switch are connected to the high-voltage contacts of the relay.

Further examination of this low-voltage, remote-control wiring system will show that it is often superior to conventional wiring, but will not replace it in every case. Low-voltage wiring, however, should be given consideration on new construction projects as well as in old work.

REVIEW QUESTIONS

1. Are batteries used with remote-control residential wiring systems?
2. Does the control relay usually have two coils?
3. What voltage is applied to the outlet-box relays?
4. May two or more outlets be controlled by a single button?
5. At what maximum distance can three or four relays be operated?
6. What is the maximum distance for operation of a single relay?
7. Does current flow through the relay coil all the time the light is turned on?
8. Do the transformers require overcurrent protection?
9. Are low-voltage and high-voltage wires intermingled?
10. Do control switches have two buttons?
11. Are outlet boxes required for mounting the switches?
12. What device permits control of relays from a central point?
13. Is No. 14 wire required for low-voltage switch loops?
14. Are three-way and four-way switches used with this system?
15. Is the relay always mounted in the outlet box?
16. How is relay noise averted?
17. Is the transformer always mounted in the panel?
18. What is the most desirable location for a selector switch?
19. How are remote-control wires usually secured?
20. What must be done if the outlet box on old work is too small to hold the relay?

Chapter 14

Outside Wiring

Electrical Power on the Farm

QUESTIONS THIS CHAPTER WILL ANSWER

1. What should be the clearance of wires crossing a road?
2. How close may uninsulated overhead conductors run to a building?
3. In general what type of wiring is used on farm property?
4. What type of cable is used for underground services on farms?
5. How is demand load calculated?

Outside Wiring

Its Importance to Inside Wiremen. Linemen who specialize in outside work are employed by utility companies. When a private operation is concerned, such men are not always available, and the interior wireman is called upon to do the job. For this reason, he should become familiar with the general principles and methods used in outdoor work.

The discussion of farm wiring, likewise, should interest the inside wireman. Points which arise in connection with it are not covered elsewhere. Also, a farm represents a city in miniature. The problem involved in determining demand load on the farm is one that applies to every type of community, regardless of size.

NEC Rules for Outside Work. Application of the NEC rules for voltages of 600 or less is illustrated in Fig. 1. The NEC provides that wires on poles shall be not less than 1' apart unless mounted on racks or brackets. Phone wires shall not be placed on the same crossarm with power wires, and shall be underneath them rather than above. Horizontal climbing space shall be maintained between

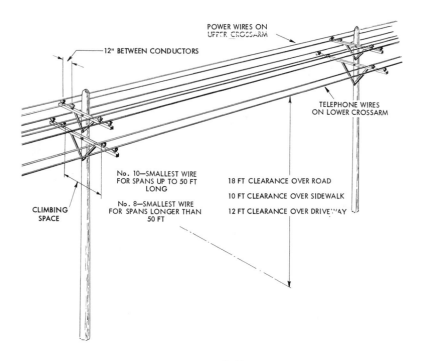

POWER WIRES ON
UPPER CROSSARM

12" BETWEEN CONDUCTORS

TELEPHONE WIRES
ON LOWER CROSSARM

No. 10—SMALLEST WIRE
FOR SPANS UP TO 50 FT
LONG

No. 8—SMALLEST WIRE
FOR SPANS LONGER THAN
50 FT

CLIMBING
SPACE

18 FT CLEARANCE OVER ROAD

10 FT CLEARANCE OVER SIDEWALK

12 FT CLEARANCE OVER DRIVEWAY

Fig. 1. Pole line

conductors on the poles, equal to 24″ between inside wires where the voltage is less than 300, or 30″ where the voltage exceeds 300.

When the span from pole to pole does not exceed 50′, the smallest allowable conductor is No. 10 AWG. If the span is greater than 50′, the smallest conductor is No. 8. Insulation on overhead wires may be rubber, thermoplastic, or weatherproof. Insulated wires passing a building or structure may not approach closer than 3′. Uninsulated conductors may not be closer to the building than 10′.

Overhead spans shall clear a road or street not less than 18′; sidewalks, 10′; an area used by automobiles or equipment, 12′. Clearance above a flat roof must be at least 8′, and above a roof which is too steep to be readily walked on, 3′. Service or feeder drops shall clear doors and windows at least 3′. Drip loops shall be formed at service heads, or where conductors enter buildings.

Lamps on poles shall be located below live conductors and transformers. It is suggested that grounding procedures, Chapter 4, be reviewed at this time. Metal enclosures of conductors shall, if

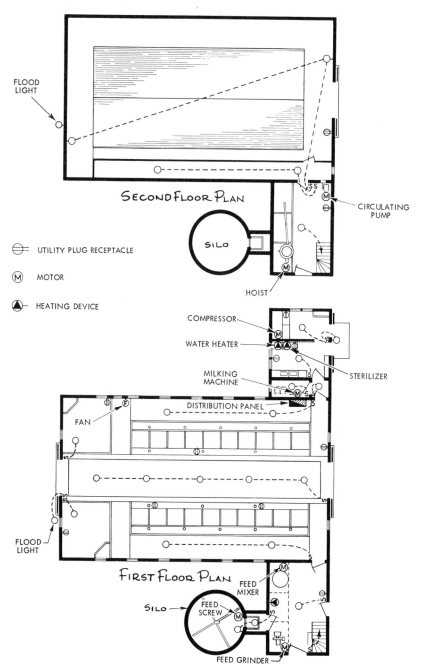

FLOOD
LIGHT

Second Floor Plan

⊖— UTILITY PLUG RECEPTACLE

Ⓜ MOTOR

◭— HEATING DEVICE

SILO

CIRCULATING
PUMP

HOIST

COMPRESSOR

WATER HEATER

STERILIZER

MILKING
MACHINE

DISTRIBUTION PANEL

FAN

FLOOD
LIGHT

First Floor Plan

FEED
MIXER

SILO

FEED
SCREW

FEED GRINDER

Fig. 2. Dairy barn wiring plan

possible, be kept at least 6' away from lightning rod conductors. Where this separation is impracticable, the pipe and the lightning rod conductor shall be bonded together.

Farm Wiring

Building. Generally, the wiring material used on farms is non-metallic sheathed cable. In damp or corrosive locations, Type UF cable may be employed. The farmhouse should be wired in exactly the same way as the bungalow studied earlier. Its connected load is assumed to be 28 kw. Service needs are calculated under the general or the optional method explained earlier, either calculation giving a result in the neighborhood of 21 kw. The current is equal to 21,000 watts divided by 230 volts, or 91.5 amps. Referring to NEC Table 310-12, the size of rubber covered type SE service entrance cable is No. 2 AWG copper.

The connected load for the dairy barn, Fig. 2, is taken at 27 kw, including 15 circuits for lights, plug receptacles, and power. Reliable test data shows that the maximum, or *demand,* load for such a location may be rated as low as 30 percent of connected load. A service installation large enough to handle about 8 kw will be deemed satisfactory here.

Where a building is exposed to highly combustible or dusty material (hayloft, silo, and the like), it is advisable to install vapor-proof fixtures. Figure 3 shows a cut-away view of a vapor- and dust-

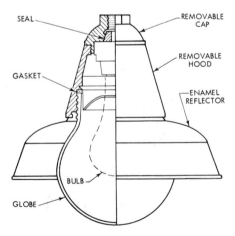

Fig. 3. Sectioned drawing of a vapor-proof lighting unit

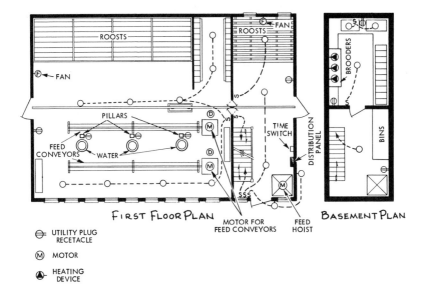

FIRST FLOOR PLAN

BASEMENT PLAN

MOTOR FOR FEED CONVEYORS   FEED HOIST

⊖ UTILITY PLUG RECETACLE

Ⓜ MOTOR

🔺 HEATING DEVICE

Fig. 4. Poultry-house wiring plan

proof lighting unit. It is possible to select these units so threaded that a fruit jar can be screwed over the lamp to protect it until another globe is obtained. Vapor-proof fixtures minimize the chances of hot broken lamps dropping onto readily combustible material. The enclosing globe prevents corrosive vapors from attacking the socket contacts. When vapor-proof fixtures are used with nonmetallic sheathed cable, means for sealing the cable entrance to the lamp must be provided.

The poultry house, Fig. 4, uses four 15-amp lighting circuits, two

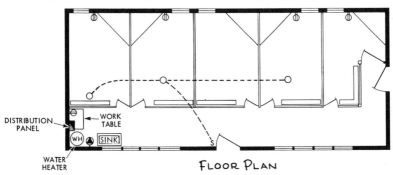

FLOOR PLAN

DISTRIBUTION PANEL

WORK TABLE

SINK

WATER HEATER

Fig. 5. Hog-house light and power drawing

of which are controlled by a time switch, three plug receptacles, a feed-machine motor, an electric-oven brooder, and a 230-volt elevator motor. The connected load amounts to 5.4 kw, but the demand load will not be greater than 4 kw.

The hog house, Fig. 5, has one lighting circuit, one plug recep-

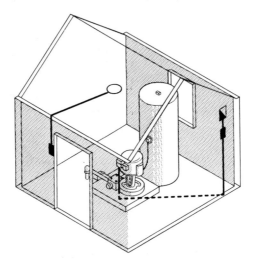

Fig. 6. Pump house light and power layout

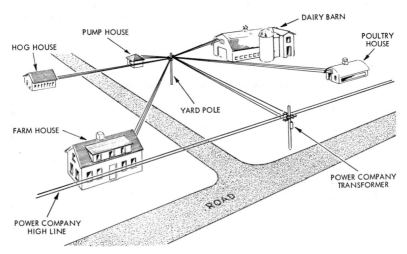

Fig. 7. Yard distribution system

tacle circuit, and a water-heater circuit. The connected load amounts to 2.8 kw; the demand load, 1.5 kw.

One of the most important buildings is the pump house, Fig. 6, because it must supply water in case of fire. For this reason, it should be located some distance from other buildings. It contains only a light and a motor. Connected and demand loads for this building are 1 kw each.

Yard Distribution Plan

A pictorial view of the distribution system is presented in Fig. 7. It shows the location of the power company installations such as high-voltage lines, transformer, feeder line to the yard pole, and lines radiating to the various buildings. The breakdown of the connected load is as follows:

> Farmhouse .... 28.0 kw
> Barn .... 27.0 kw
> Poultry House .... 5.4 kw
> Hog House .... 2.8 kw
> Pump House .... 1.0 kw
>
> ———————
>
> Total connected load . . 64.2 kw

On the basis of reliable test data, demand load of a farming cen-

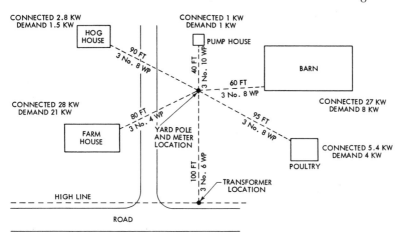

SCALE: 1/4" = 20'-0"

Fig. 8. Line diagram of distribution system

ter may be taken as 30 percent of total connected load, including that of the farmhouse. The demand load of the farming center in this instance is equal to 0.3 × 64.2 kw, or 19.3 kw.

A line diagram of the yard distribution system should be constructed as in Fig. 8, distances and conductors marked in, and wire sizes checked against permissible voltage drop. It should be noted that weatherproof conductors are used here.

Current in the main line from the power company transformer to the yard pole is equal to 19,300 watts divided by 230 volts, or 84 amps. NEC Table 310-13 (reprinted in the appendix) lists current-carrying capacity of insulated conductors in free air. The values given in the right-hand column, under Type SB, apply to weather-proof wire as well. The rating of No. 6 AWG conductor is 100 amps.

The voltage drop in 200′ (twice the one-way distance) of No. 6 equals 84 amps × R, where R is the resistance of the wire. Using the formula of Chapter 2, $R = \dfrac{10.8 \times 200}{26{,}250}$, or approximately 0.08 ohm. Volts drop = 84 amps × .08 ohm = 6.72 volts. Under balanced load conditions, where each line wire carries exactly the same amount of current and the neutral carries no current, the voltage between either line wire and neutral is 115 volts — 3.36 volts, or 111.64 volts. This drop is somewhat high, but the power loss is absorbed by the power company, because it occurs ahead of the customer's meter. The company will, no doubt, install No. 6 AWG wire.

The farmhouse load is 21 kw, and the current 91.5 amps. Here, too, No. 6 may be used. The voltage drop for 160′ of this conductor is equal to 91.5 amps × .064 ohm (the resistance of 160′ of wire), or 5.86 volts, which is too high. If No. 4 wire is substituted, with a resistance of 0.04 ohm for 160′, the voltage drop is reduced to: 91.5 amps × .04 ohm, or 3.68 volts, which is acceptable.

A No. 10 AWG wire has sufficient carrying capacity to handle the 8-kw barn load, but since the span is over 50′, a No. 8 AWG conductor must be employed. The resistance of 120′ of No. 10 conductor is 0.12 ohm. The current represented by 8 kw is approximately 35 amps. The voltage drop equals 35 amps × .12 ohm, or 4.2 volts, which is satisfactory for a load of this kind.

In the poultry house, which spans 95′, No. 8 AWG conductors

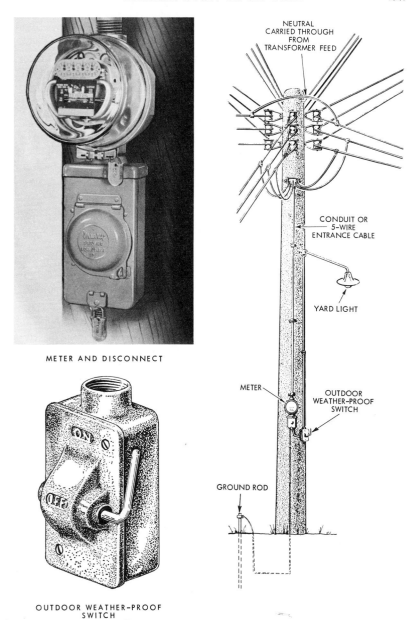

METER AND DISCONNECT

OUTDOOR WEATHER-PROOF SWITCH

NEUTRAL CARRIED THROUGH FROM TRANSFORMER FEED

CONDUIT OR 5-WIRE ENTRANCE CABLE

YARD LIGHT

METER

OUTDOOR WEATHER-PROOF SWITCH

GROUND ROD

Fig. 9. Typical yard pole meter and disconnect

Courtesy of Murray Mfg. Corp.

must be employed, as in the hog house. To the pump house, which is only 40′ from the yard pole, No. 10 conductors may be used. From the line-drop calculations performed in connection with the other loads, it is evident that the voltage drop to each of these three locations will be acceptable.

Yard Pole. A single yard pole is shown in Figs. 7 and 8, but a particular installation may require more than one. In deciding upon the number of poles, the following factors should be considered:

(1) Sag in wires due to expansion of metal in warm weather.
(2) Stretching caused by weight of sleet, ice, and snow during the winter season.
(3) Inadvisability of drawing line wires too tight in the process of installing them.

The yard pole, Fig. 9, has six 3-wire brackets which support the lines from the transformer and the five sets of drops to the various buildings. The neutral supply wire is connected to all five sets of drops at the top of the pole, and a tap from the neutral is carried down to the meter and service switch as a neutral grounding conductor. This wire is attached to the service switch enclosure, and continues downward to a driven ground rod at the foot of the pole. The yard light is controlled by a weatherproof switch, the metallic enclosure of which is also grounded. A radial system of feeding the buildings is not always feasible. Where buildings are arranged in a line, it may become necessary to employ a series of poles or racks to aid in distribution.

Underground Wiring. Underground wiring systems have certain advantages even though they are more costly to install than equivalent overhead systems. They are more permanent, and not so readily damaged. Type USE cable can be placed directly in the ground. It should be protected by metal conduit to a height of 8′ where the cable emerges from the ground to run up a pole, and should enter a building through a conduit elbow that projects through the foundation wall.

REVIEW QUESTIONS

1. What types of insulation are approved for overhead wires?
2. What separation must be maintained between span wires on crossarms?

3. What is the minimum width of climbing space for voltages of 300 or less?

4. What is the minimum width of climbing space for voltages over 300?

5. What is the minimum clearance of span wires above a road?

6. State the minimum clearance above a sidewalk.

7. State the minimum clearance above a lot where machinery is employed.

8. How close may uninsulated span wires approach a building?

9. How close may insulated span wires approach a building?

10. What is the minimum clearance of span wires above a flat roof?

11. How close may service or feeder drops approach a window?

12. Should pole-mounted lighting fixtures be placed above or below conductors?

13. How is yard pole service equipment grounded?

14. Why should the pump house be located away from other buildings?

15. What is meant by the term *demand load?*

16. What percentage expresses the demand load of a farming community?

17. What type of wiring is commonly employed on farm buildings?

18. What type of cable should be used in damp places?

19. What type of cable may be used for underground services?

20. What particular factor makes voltage drop such an important consideration with respect to farm loads?

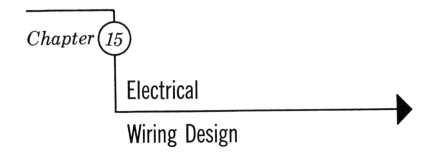

Chapter 15

Electrical

Wiring Design

QUESTIONS THIS CHAPTER WILL ANSWER

1. How should residential lighting circuits be distributed?

2. What special provision must be made with respect to the kitchen, dining room, and laundry of a residence?

3. How is service load determined in the average small dwelling?

4. How is service load determined under the optional method?

5. What are the smallest sizes of Type R or Type TW wire required by the National Electric Code for general lighting, utility, and service circuits?

GENERAL METHOD

Basic NEC Requirements

The code requires that a load of 3 watts be provided for each square foot of living quarters in residential occupancies. It also recommends that a circuit be provided for each 500 sq ft. In determining this area, the outside dimensions of the building shall be used.

All lighting and plug receptacle outlets (convenience outlets) may be connected to the circuit which takes in a 500 sq ft space, except in kitchens, dining room, laundry, and similar locations. In the latter rooms, plug receptacles may not be connected on lighting circuits, but on two or more 20-amp utility circuits.

The total area of this bungalow, using outside dimensions, is 28 ft × 55 ft, or 1540 sq ft. Dividing by 500 sq ft, it is seen that slightly more than three lighting circuits will be needed. In such case, four circuits should be employed, even though inspection

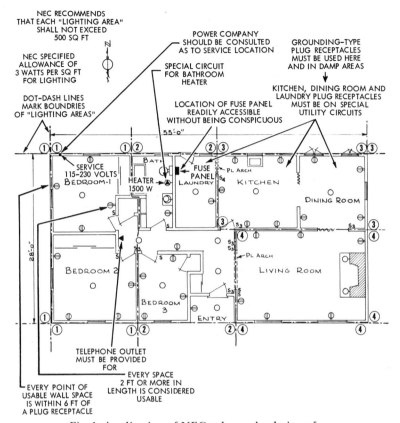

Fig. 1. Application of NEC rules to the design of
residential wiring installation

jurisdictions, in some instances, permit a small overage in making the calculation.

The dot-dash lines, in Fig. 1, bound lighting areas. The Code recommends a limit of 500 sq ft for them. Here, one such space is designated as (1), the second as (2), the third as (3), and the fourth as (4). The area of the first is approximately 413 sq ft, the second 455 sq ft, the third 336 sq ft, and the fourth 345 sq ft. Judgment must be exercised in assigning coverage for each circuit so that the wiring layout may be as simple as possible.

Calculation of Load

The minimum lighting power, under NEC rules, is equal to

1540 × 3 watts, or 4620 watts. The two utility circuits must be taken at 1500 watts each, or 3000 watts for both. The NEC permits the lighting and utility wattages to be added together before applying reduction percentages or demand factors, allowed under Table 220-4(a). The total here is 4620 watts + 3000 watts, or 7620 watts.

NEC Table 220-4(a), which is reprinted in the Appendix, states that the first 3000 watts of such load must be assessed at 100 percent, and that the remainder, up to 120,000 watts, must be calculated at 35 percent. Applying these rules:

First 3000 watts	= 3000 watts
Remainder (7620 watts minus 3000 watts) equals 4620 watts.	
0.35 × 4620 watts	= 1617 watts
	————
Load for lighting and utility circuits	= 4617 watts

In addition to these circuits, there is a bathroom heater, which must be calculated, under the NEC, at full load of 1500 watts.

The total service load, then, is the sum of 4617 watts and 1500 watts, or 6117 watts.

The supply is 3-wire, 115-230 volts, and the current equals 6117 watts divided by 230 volts, or 27 amps. NEC Table 310-12 shows that a No. 10 Type R or Type TW conductor has sufficient current-carrying capacity for this amperage. The NEC provides, however, that where more than two circuits are involved, the smallest service conductor shall be No. 6 AWG. The service switch must be not smaller than 60 amp. Should a circuit-breaker be used as the service disconnecting means, its rating need not be greater than 50 amp unless the load current exceeds this value.

A No. 10 AWG feeder conductor from service switch to distribution panel would be acceptable. This conductor, however, would not furnish extra current-carrying capacity over that required to supply seven circuits. Three-wire, single-phase, distribution panels are made with even numbers of circuits, eight being included here. To provide additional capacity so that the extra circuit may be utilized at some future date, it is advisable to employ a No. 8 AWG feeder instead. The difference in material cost is only about three dollars, while that of labor is nominal.

Additional Loads.

If a 5000-watt, 230-volt electric clothes dryer is to be installed, its current input is equal to 5000 watts divided by 230 volts, or approximately 22 amps. When this current is added to the 27 amps calculated above, the total service load becomes 27 amps + 22 amps, or 49 amps. Table 310-12 (NEC) shows that the current-carrying capacity of the No. 6 service conductor (55 amp) has not been exceeded. Therefore, the present service conductor and switch are large enough to carry the dryer in addition to the original load.

A 1/6 hp, 115-volt, garbage disposal unit could also be installed without altering the present service equipment. It consumes about 500 watts, the current being 4.35 amps. Each of the outer service wires carried 49 amps before the garbage disposal unit was connected. One of them will now carry 49 amps + 4.35 amps, or 53.35 amps, a value still within the current-carrying capacity of the No. 6 conductors and the 60-amp service switch. It is worth noting, however, that the current-carrying capacity of a 50-amp circuit breaker will be exceeded.

The inclusion of a 1500-watt, 115-volt dishwasher will call for increasing the size of service conductors and of the service disconnect switch. The current taken by this unit is equal to 1500 watts divided by 115 volts, or approximately 13 amps. Since the dishwater will be connected to the opposite line wire from the garbage disposal unit, the highest current in this wire must be equal to: 49 amps + 13 amps, or 62 amps, requiring a No. 4 Type R or Type TW copper conductor and a 100-amp switch.

Consider now, what happens if the owner purchases a 12-kw electric range. NEC Table 220-5 allows a demand factor to be employed in connection with cooking appliances. A 12-kw range is assessed, in column 2 of Table 220-5, (see Appendix), at 8 kw. The current is equal to 8000 watts divided by 230 volts, or approximately 35 amps. Adding this amount to the 62 amp present load, gives: 62 amps + 35 amps, or 97 amps. Table 310-12 lists a suitable Type R or Type TW conductor as AWG No. 1. The service conductors must be changed, in this case, but the 100-amp service switch is still adequate.

As a comparison between an electric range and two separate units of oven and cooking top, suppose that the oven has a rating of 4.5 kw and the cooking top 7.5 kw, the total connected load being 12 kw, as before. Referring again to NEC Table 220-5, Column A permits two units, not over 12 kw each, to be assessed at 11 kw. Column C, on the other hand, allows a demand factor of 65 percent for two units whose kw ratings are between 3.5 and 8.75 kw.

The devices qualify under this condition and, since their combined input is 12 kw, their load may be taken at 65 percent of this value, or .65 × 12 kw, which equals 7.8 kw. The current taken by them is equal to 7800 watts divided by 230 volts, or approximately 34 amps, requiring the same size conductor as in the case of the range. Comparative results do not always turn out so close as in the present instance, but they are usually in the same general neighborhood.

Optional Method

Assume that, in addition to the 97-amp requirement in the above example, there is a 5-kw air-conditioning unit and an 8-kw central heating plant. Where there are two such opposite types of load, both of which are not used at the same time, the NEC permits omission of the smaller one in making service calculations.

In the present example, the 5-kw air-conditioning unit may be ignored, but the 8-kw heating plant must be considered. The current taken by the central plant is equal to 8000 watts divided by 230 volts, or approximately 35 amps. The total wattage is now equal to 97 amps + 35 amps, or 132 amps. NEC Table 310-12 shows that a No. 00 Type R or Type TW conductor will be needed in addition to a 200-amp service switch.

The NEC permits an optional method for calculating load in a single-family residence which is served by a 115-230 volt, 3-wire, 100-amp (or larger) service. The method is set forth in NEC Table 220-7, (see Appendix).

The Code requires that air-conditioning or space heating loads be calculated at 100 percent, but that "all other loads" be granted a demand factor. The demand factor is applied after all other load, including lighting, utility, range, dryer, dishwasher, and similar loads are first assessed at 100 percent value. The method is illustrated in the following example:

Lighting load	1540 × 3 watts =	4620 watts
Utility load	2 × 1500 watts =	3000 watts
Electric range (full-load nameplate rating) =		12000 watts
Clothes Dryer	=	5000 watts
Dishwasher	=	1500 watts
Garbage disposal unit	=	500 watts
Total of "all other load"	=	26620 watts

The assessed value of this load is equal to:

First 10000 watts	=	10000 watts
Remainder (26620 watts — 10000 watts)		
.4 × 16620 watts	=	6648 watts
Total accepted value of load	=	16648 watts

The heating load is equal to 1500 watts (bathroom heater) + 8000 watts (central plant), or 9500 watts.

The total input is equal to 16648 watts + 9500 watts, or 26148 watts. The current necessary to supply this power is equal to 26148 watts divided by 230 volts, or approximately 114 amps. Table 310-12 (NEC) shows that a No. 0 Type R or Type TW copper conductor is large enough for this current. As a result, a saving of one size in conductor rating is effected by use of the optional method in this particular instance.

Other Types of Occupancy

Wiring calculations for other residential occupancies, such as apartment houses and hotels, are performed in the same general way as for the single-family dwelling. Identical demand factors are applied to standard apartment house units. In hotels, motels, and those apartment houses which have no provision for cooking by tenants, the demand factor percentages of NEC Table 220-4(a) are seen to be different. The NEC provides, also, for a 75% demand factor in all types of dwellings where four or more fixed appliances other than electric ranges, air-conditioning equipment, or space-heating equipment, are connected to the same feeder or service wires.

REVIEW QUESTIONS

1. What service capacity does NEC require for each square foot of residential occupancy?
2. What maximum area should be supplied by one lighting circuit?
3. What is the ampere rating of each utility circuit?
4. How many utility circuits are required by the NEC?
5. May a hall plug receptacle be connected to the utility circuit?
6. Are demand factors applied to both lighting and utility circuits under the NEC?
7. How much wattage must be assigned to two fully-loaded lighting circuits and two utility circuits?
8. What demand factor is granted to a bathroom heater under the NEC?

9. What size of service conductor should be employed for six lighting circuits which are fully loaded, and two utility circuits?

10. What service switch is needed for the above installation?

11. What is the smallest circuit breaker that could be used here?

12. What load must be assigned to a 5000 watt clothes dryer?

13. How much current must be allowed for, under the general method, to accommodate a 12-kw, 115-230-volt range?

14. How much current must be provided for 115-230-volt oven and cooking top whose total load equals 11 kw?

15. Is a bathroom heater included in "other load" under the optional method of service calculation?

16. Under the optional method, how much load is assigned to a 12-kw range during the process of making calculations?

17. What demand factor is granted an air-conditioning load under the optional method?

18. How much load is calculated for an 8-kw heating load under the optional method if it is smaller than the air-conditioning load?

19. Are lighting demand factors the same for all types of residential occupancies?

20. What demand factor is allowed, under the NEC, where four or more fixed appliances, other than ranges, air-conditioners, or space heaters, are connected to a service?

Chapter 16

Estimating
Electrical Wiring

QUESTIONS THIS CHAPTER WILL ANSWER

1. *What are the essential factors that enter into an estimate?*
2. *What is a branch-circuit material schedule?*
3. *How are labor units determined?*
4. *How is the quantity of wire or cable ascertained?*
5. *What facts are noted on the final estimating form?*

Introduction

The term *estimate* means an opinion, formed in advance, of the probable cost of a certain piece of work. The opinion is arrived at after considering each factor, including cost of material, cost of labor, overhead cost, and expected profit. It should be emphasized that an estimate is not an exact figure, but an approximation. The precise cost may be determined only after the job has been finished.

On large construction projects, detailed drawings and specifications are made available to the estimator. On lesser operations, such as that of wiring a bungalow, drawings seldom present complete electrical layouts, while specifications are usually contained in one or two paragraphs of a letter. In the present instance, the electrical drawing is reasonably complete, except that the conduit run to the telephone outlet in the hall and the whole bell-wiring system are omitted in order to simplify matters.

Specification paragraphs in the "contract" letter set forth certain requirements which may be summarized as follows: All electrical material shall bear U-L approval. Workmanship shall be of high quality. The distribution panel shall be of the circuit-breaker type.

233

Cable shall be fastened by means of clamps in all outlet boxes. A weather-tight cover shall be installed on the plug receptacle outlet in the entry. A telephone outlet shall be installed in the hallway. The owner's approval shall be obtained in the selection of lighting fixtures which, if necessary, shall be in accordance with a list to be supplied by him upon request. Bell system chimes and the front door push button must have the owner's approval. Material and workmanship shall be guaranteed against inherent defects arising within a period of one year from date of completion. The completed installation shall conform to the NEC, and shall meet with the approval of local inspection authorities.

Preliminary Steps

The essential requirement in every type of estimating is to follow a systematic procedure. The first consideration is to become familiar with the plan, Fig. 1, noting the general layout of rooms, location of electrical service and telephone service. Special outlets are looked for, such as the heater in the bathroom. Location of the distribution panel is checked.

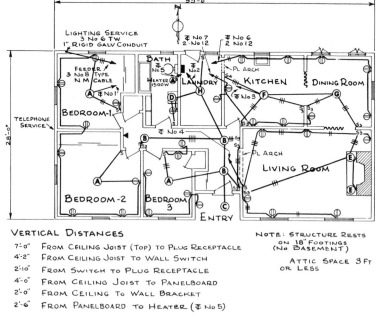

VERTICAL DISTANCES

7'-0"   FROM CEILING JOIST (TOP) TO PLUG RECEPTACLE
4'-2"   FROM CEILING JOIST TO WALL SWITCH
2'-10" FROM SWITCH TO PLUG RECEPTACLE
4'-0"   FROM CEILING JOIST TO PANELBOARD
2'-0"   FROM CEILING TO WALL BRACKET
2'-6"   FROM PANELBOARD TO HEATER (₵ No 5)

NOTE: STRUCTURE RESTS ON 18" FOOTINGS (No BASEMENT)

ATTIC SPACE 3 FT OR LESS

Fig. 1. Drawing from which estimate is prepared

The occupancy is a one-story bungalow, having a low attic and no basement. These facts have an important bearing on the manner of running cable between certain outlets. The subfloor will be in place when the electrician arrives, so that he will not be able to run cable along sides of floor joists. At door openings, he will find it necessary to cross by way of the attic.

In the living room, for example, it will not be possible to route cable directly across from the first plug receptacle outlet in the north wall to the switch location at the side of the archway. Instead, the cable must be carried upward to the ceiling from the receptacle outlet, across tops of ceiling joists, and down again to the switch location. Since the attic is unsuited for living quarters, the cable will not require protection where it is run exposed. Identical situations are found throughout the premises.

Branch Circuit and Fixture Schedules. Having made a general survey of the plan, it is necessary to draw up a branch-circuit schedule, Fig. 2. The form lists ceiling and bracket outlets, single-pole, three-way, and four-way switches, standard plug receptacles, and grounding-type receptacles. The last column deals with special outlets.

BRANCH CIRCUIT SCHEDULE									
		LIGHTING OUTLETS		SWITCH OUTLETS			RECEPTACLE OUTLETS		SPECIAL
CIRCUIT	LOCATION	CEILING	BRACKET	SP	3-W	4-W	O	G	
1	BEDROOM-1	1		1			4		
	BEDROOM-2	1		1			4		
2	BATHROOM		1	1			1		
	LAUNDRY	1		1					
	BEDROOM-3	1		1			4		
	HALL AND ENTRY	4			4		1	1	
3	KITCHEN	1			2	1			
	DINING ROOM	1			2				
4	LIVING ROOM		2	1	2		6		
5	BATHROOM								1
6	KITCHEN							2	
	DINING ROOM							3	
7	KITCHEN							3	
	DINING ROOM							1	
	LAUNDRY							1	

Fig. 2. Branch circuit schedule

Circuit *1* supplies Bedrooms 1 and 2, having a ceiling outlet, single-pole wall switch, and four convenience outlets in each one. Circuit *2* takes care of lighting and plug receptacle outlets in bathroom, Bedroom 3, hall and entry. The laundry ceiling light is connected to it, but the laundry plug receptacle is not.

In kitchen and dining rooms, lighting outlets, but not plug receptacles, are connected to Circuit *3*. Circuit *4* supplies both lighting outlets and plug receptacles in the living room. Circuit *5* has one special outlet connected to it—the bathroom heater. Circuits *6* and *7* are reserved for plug receptacles in kitchen and dining rooms. The laundry plug is also connected to Circuit *7*.

CIRCUIT	TYPE							
	A	B	C	D	E	F	G	H
1	2							
2	1	3	1	1				1
3						1	1	
4					2			

FIXTURE SCHEDULE

Fig. 3. Lighting fixture schedule

The fixture schedule, Fig. 3, is drawn up from type letters marked on the various outlets shown on the drawing. Their exact specifications are to be determined through consultation with the owner. There are three type *A* units, three type *B* units, two type *E* units, and one unit each of types *C, D, F, G,* and *H.*

Branch Circuit Material Schedules. Drafting of a branch-circuit material schedule, Fig. 4, becomes the next order of business. It is constructed on the basis of measuring and counting. Runs of cable from outlet to outlet are measured on the plan. Other items, such as outlet boxes, plaster rings, switches, plug receptacles, plates, box hangers, ground straps, and wire nuts, are counted. The measuring can be done with the aid of a rotameter or an architect's scale, the counting by means of a metal tabulator, or by mental arithmetic. The mechanical aids are reserved for large commercial or industrial jobs. Here, the architect's scale and mental arithmetic will be employed.

Before starting to measure cable, it is well to note the table

CIRCUIT	CABLE			BOXES				PLASTER RINGS						SWITCHES			PLUG RECEPTACLE		PLATES				WIRE NUTS	CONNECTORS	HANGERS	GRD STRAPS
	14-2	14-3	12-2	4"O	4"□	4½	SR	4"R	4"SQ	4"SS	4"2G	4½R	4½S	SP	3-W	4-W	ST	GR	SS	2-G	PR	SP				
1	129'-0"	10'-8"		2			10	2						2			8		2		8		4	2		
2	152'-0"	35'-2"		4	3	1	12	4	2			1	1	3	4		5	2	5	1	6	1	14	2	7	2
3	53'-4"	44'-8"			1	1	5	1				1			4	1			5				4	2	2	1
4	78'-8"	76'-10"		1	3	1	5		2	2	1			1	2		6		1	1	6		4	2	2	
5	4'-6"																									
6			54'-0"				5											5			5					
7			54'-0"				5											5			5					
SUB-TOTAL	417'-6"	167'-6"	108'-0"																							
ADD 10% TO CABLE	42'-0"	17'-0"	11'-0"																							
TOTAL	459'-0"	184'-0"	119'-0"	7	7	3	42	7	4	2	1	2	1	6	10	1	19	12	13	2	30	1	26	6	13	3

BRANCH CIRCUIT MATERIAL SCHEDULE

(arrow to cable TOTAL figures) — ROUND FIGURES

Fig. 4. Branch circuit material schedule

of vertical distances at the bottom of the drawing. A 7′ indication marks the distance from the top of a ceiling joist down to a plug receptacle. A 4′-2″ line shows the distance from top of ceiling joist down to a wall switch. The distance from a switch down to a plug receptacle is shown as 2′-10″; that from ceiling joist to top of the distribution panel is 4′; and that from ceiling joist to wall bracket is 2′. The last dimension, 2′-6″, is from panelboard down to the heater location (Circuit 5).

Circuit 1 ends at the wall switch in Bedroom 2, passing through the various plug receptacle, switch, and ceiling outlets in both rooms, and then to the distribution panel in the laundry. The cable is No. 14-2 throughout, except for a short run of No. 14-3 in Bedroom 1, from the ceiling light to the wall switch. Distances to ceiling outlets are measured to their centers, while distances to plug receptacles and to switches are measured to the wall. Do not try to estimate fractional amounts. If the measurement is slightly beyond a given mark on the scale, call it 6″; if slightly less than a given mark, grant it the whole 1′.

Starting at the end of the circuit, 4′-2″ of cable must be allowed for the distance upward from the wall switch to the top of a ceiling joist, and 7′-6″ for the run across the joists to the light in the center of the room. Another 7'-6" to the right marks a point directly above the plug receptacle on the east wall of the room. From this point, 7' are added to account for the vertical distance from ceiling joist down to the plug receptacle.

Continuing, the distances are 6' from the plug receptacle to the south wall of the room; 14'-6" to the west wall; 14' to the north wall; 5' to the plug receptacle in the south wall of Bedroom 1; 5' back to the west wall; 13' to the north wall; 6'-6" to the plug receptacle in the north wall; 7' vertically to the top of the ceiling joist; and 6'-6" to the ceiling outlet.

From this point, the measurement to the distribution panel in the laundry is 17', but 4' must be added to provide for the run from the ceiling joist down to the panel. Another short piece of No. 14-2 cable extends from the wall switch in Bedroom 1 down to the plug receptacle, the length being equal to the sum of 2'-10" and 1'-6", or 4'-4". The quantity of No. 14-3 cable in Bedroom 1 is equal to 6'-6" from center light to the wall, plus 4'-2" from ceiling joist down to the switch, a total of 10'-8".

Adding these amounts, the length of No. 14-2 cable is found to be 129', while that of No. 14-3 cable is 10'-8". It may be observed that no allowance was made for the short pieces of cable which project from each outlet box. This matter is taken care of by the addition of 10 percent to the measured lengths, as will be shown later on. For the present, a notation of 129' is made in the 14-2 column, and 10'-8" in the 14-3 column of the schedule. Cable lengths for the remaining circuits are then entered in appropriate columns, including the No. 12-2 conductors of the utility circuits.

Other materials are now taken off. The method will be illustrated with respect to Circuit *1*. Center lights in the bedrooms require two 4" octagon boxes, two box hangers, two 4" round plaster rings, and four wire nuts that will be used for connecting the fixtures. Ten switch-receptacle boxes are needed, eight for the plug receptacles and two for the wall switches. Two single-pole switches will be required, and two single switch plates. Eight standard plug receptacles will need eight receptacle plates. These amounts are noted in appropriate columns of the schedule.

Remaining circuits are examined in the same way. Circuit *2* is somewhat more complicated in its requirements than Circuit *1*. Before considering Circuit *2*, it should be explained that the NEC limits the number of conductors that may enter a given size of outlet box. A 4″ standard octagon box may be used for not more than eight No. 14 conductors, or for seven No. 14 conductors where the box contains cable clamps. A 4″ square box will handle eleven No. 14's, or ten No. 14's where cable clamps are present. A $4\frac{11}{16}$″ box will handle sixteen No. 14 conductors, or fifteen where cable clamps are present. The number of wires in the cable runs are indicated by cross marks on the cable. Where there are no cross marks, it is understood that two wires are run.

The ceiling outlet box in the laundry has four cables, none of which has cross marks. The total number of wires, then, is equal to 2x4, or 8. Since cable clamps are present, a 4″ square box must be used. An identical situation is found with respect to the hall lighting outlet which is just inside the front door. Eleven No. 14 wires enter the hall ceiling outlet which is nearest the kitchen door. Since it has cable clamps a $4\frac{11}{16}$″ outlet box is needed. The other ceiling outlets take 4″ octagon boxes. A 4″ square box is needed for the two three-way switches that are ganged, or grouped, at the side of the living-room archway.

Four 4″ round plaster rings are necessary for the octagon boxes, two 4″ square plaster rings for the square boxes, one two-gang 4″ plaster ring for the ganged switches, and one $4\frac{11}{16}$″ square plaster ring for the hall ceiling outlet. Three single-pole and four three-way switches are needed, five standard plug receptacles and two of the grounding type, five single switch plates, one two-gang switch plate, six standard plug receptacle plates and one special weatherproof cover for the entry receptacle, fourteen wire nuts, two 1″ box connectors for the $4\frac{11}{16}$″ box, seven box hangers, and two ground straps.

Circuits 3, 5, and 7 offer no difficulties in the listing of material, but there are some points worth noting in connection with Circuit 4. A $4\frac{11}{16}$″ box is required for the two-gang switch outlet just inside the living-room arch, because there are eleven wires in addition to cable clamps. Three 4″ square boxes are used on this circuit, one for the north bracket outlet at the fireplace, and the others for two plug receptacle outlets. The reason for the square box at the fireplace is readily apparent. A 4″ box is substituted for the standard receptacle box in the two outlets adjacent to either side of the hall archway because six wires are present at these locations. It is good practice to use the 4″ square boxes in place of ordinary switch-receptacle boxes whenever the number of wires exceeds five. After

SERVICE AND FEEDER MATERIAL SCHEDULE		
ITEM	NUMBER	FEET
1" ENTRANCE CAP	1	
1" GALV CONDUIT		10
1" GALV LOCKNUT	2	
1" GRD BUSHING	1	
1" GALV STRAPS	3	
1" METER SOCKET	1	
SERVICE SWITCH	1	
GROUND CLAMP	1	
No. 6 TYPE RW		36 *
1" NIPPLE	1	
SERVICE GROUND WIRE		10
DIST PANEL	1	
No. 8 N M CABLE		34 *
CABLE CLAMP	5	
40-A FUSE	2	
1¼"- 10 WOOD SCREWS	12	
* 10% NOT INCLUDED		

Fig. 5. Service and feeder material schedule

a ten percent margin has been added to measured wire lengths, the columns are totalled.

Other Method of Take-Off. The method used here may be varied to suit individual tastes. Some estimators prefer to list wire for the whole area at one time, and to count all the various items as well. After gaining experience, one may proceed in this way. At the start, however, it is well to list materials circuit by circuit in order to recheck uncertain portions without too much loss of time.

In any event, branch circuit material schedules follow the general pattern set forth here. When the take-off is done on an area basis, separate counts are made by floors, sections, or buildings, the various locations being noted, for example, as: "Floor 1," "Floor 2," and so on, instead of "Circuit *1*," "Circuit *2*," and so on.

Service and Feeder Material Schedule. This schedule may be equally as involved as the branch circuit material schedule on certain types of jobs. In the present instance, however, it is much simpler. In Fig. 5, items are listed vertically rather than horizontally. Each kind of material is enumerated. A note calls attention to the fact that an allowance of ten percent has been added to both service conductors and to feeder cable.

Labor-Unit Schedule. A table of labor units is given in Fig. 6. It includes every electrical task which enters into the complete wiring job. The right-hand column gives the time unit for performing a particular task. The units normally employed should be largely the result of one's personal experience. Trade associations publish tables which vary in degree of accuracy. When they are used by the contractor, their suitability should be checked from time to time on the basis of actual job records. Meanwhile, the present list will serve as a guide here.

The first notation states a time of 0.4 hr (24 minutes) to install a hundred feet of No. 14-2, No. 14-3, or No. 12-2 cable. This period includes boring of holes with an electric drill, pulling cable through holes or across ceiling joists, cutting the length of cable at each outlet, and stapling it where necessary.

The time for installing the No. 8-3 cable is given as 1.5 hr.

LABOR UNIT SCHEDULE	
OPERATION	**HOURS**
INSTALL 14-2, 14-3, 12-2 CABLE	.4 C FT
INSTALL 8-3 CABLE	1.5 EA
MOUNT OCTAGONAL BOX, FASTEN CABLE	.2 EA
MOUNT SWITCH BOX, FASTEN CABLE	.2 EA
ATTACH PLASTER RING	.05 EA
CONNECT RECEPTACLE, INSTALL PLATE	.2 EA
CONNECT SP SWITCH, INSTALL PLATE	.2 EA
CONNECT 3-W SWITCH, INSTALL PLATE	.3 EA
CONNECT 4-W SWITCH, INSTALL PLATE	.4 EA
SOLDER AND TAPE SPLICE	.07 EA
MOUNT AND CONNECT PANEL	2.5 EA
MOUNT AND CONNECT SERVICE SWITCH	1.9 EA
INSTALL 1" RIGID CONDUIT	9.0 EA
PULL IN No. 6 WIRE	2.0 C FT
CUT AND THREAD 1" CONDUIT	.5 EA
INSTALL METER SOCKET	1.0 EA
INSTALL ENTRANCE CAP	.3 EA
INSTALL SERVICE GROUNDS	1.0 EA
INSTALL A OR F FIXTURE	.5 EA
INSTALL B, C, D OR E FIXTURE	.4 EA
INSTALL G FIXTURE	.7 EA
INSTALL H FIXTURE	.3 EA
INSTALL ½" EMT (TELEPHONE)	4.5 C FT
INSTALL TELEPHONE OUTLET BOX	.1 EA
INSTALL BELL WIRING	1.5 EA
INSTALL "LOCAL" GROUNDS	.05 EA

NOTE: TOTAL OF 13 LOCAL GROUNDS
INCLUDING 12 PLUG RECEPTACLES
AND BATHROOM BRACKET OUTLET

Fig. 6. Labor-unit schedule

The designation "Ea" (abbreviation for "each") shows that this amount is for doing the whole job. The reason for making a special or "lump" allotment instead of supplying the 100-ft rate, is that only a short length of cable (34 ft) is involved, and only two holes need be drilled. Imposition of the general rate would, therefore, be misleading. In such case it is advisable to estimate time for the whole operation.

The unit of 0.2 hr (12 minutes) for mounting an octagon box includes assembling box and hanger, spotting the outlet, nailing the hanger to the joist, and tightening cable clamps. The time for soldering and taping a splice is 0.07 hr (about 4 minutes), which covers stripping of outer wrapping, skinning the conductors, twisting them together, soldering the joint, and covering it with plastic tape.

Service operations present some points worth mentioning. The operation of cutting and threading the 1″ service conduit is given a special rating of 0.5 hr (30 minutes). If a number of pipes were to be threaded, the time for each would be somewhat less than this value. Here, it is necessary to unpack equipment and set up the pipe vise for a single cut.

The item *install service grounds* embraces connection of a bond wire to the service grounding bushing, installation of the armored ground wire, fastening the service ground wire to the ground clamp, and fastening the ground clamp to the waterpipe grounding electrode. The time also covers attachment of the feeder ground wire to the service bonding conductor. Here, the feeder cable has a bare fourth wire that carries the service ground to the distribution center, at which point it must be connected to the metal enclosure of the panel.

Time for mounting the panel includes fastening it to the wall, inserting and tightening cable clamps, skinning feeder conductors and attaching them to the buses, grounding the enclosure by means of the bare ground wire in the cable, skinning circuit conductors, and securing them to the circuit breakers. Finally, the panel cover is set in place.

A special time rating is given to bell wiring, a value based upon numerous similar installations. The last item is that of installing local grounds. The term *local grounds* means those located elsewhere than at the service. They are commonly known as *equipment grounds.* The designation covers grounding of the bracket light in the bathroom, which is directly above the lavatory, and within reach of one who has a hand on the metal water faucet. The bathroom plug receptacle is grounded, as is the entry plug receptacle. A grounding wire is also necessary for the grounding type plug receptacles on the utility circuits. The wire is looped from box to box, and fastened under a convenient screw, such as that of a cable clamp. The end is attached to a copper ground strap which is wrapped around a water pipe. Three ground straps are used on this job.

Estimating Form. When the schedules are completed, information they contain is transferred to an estimating form, Fig. 7. Column 2, under the heading *Description,* lists every item in the schedules, as well as a number of miscellaneous components including solder, soldering paste, insulating tape, nails, bell wiring material, grounding wire, straps, and materials required for the telephone outlet.

From the information in the *Number* and *Feet* columns, together with unit costs given in Column 5, it is easy to determine material costs in Column 6. Labor units from Fig. 6 are copied, Column 7, beside appropriate items. The number of hours, Column 8, is determined by multiplying the labor unit by quantity of goods used. Finally, labor cost is found in Column 9, hours of labor being multiplied by the labor rate of $3.50 per hour.

An example will show how the estimating form is used. Item 1 is given as No. 14-2 Type NM cable. Column 4 shows that 459 ft are used. Column 5 lists the cost as $4.07 per hundred feet, or 4.59 x $4.07, which equals $18.68, as per Column 6. The labor unit for installing 100 ft of this cable is given as 0.4 hr in column 7. The time for running 459 ft of cable is equal to 4.59 x .4 hr, or 1.84 hr, column 8. The labor cost is equal to 1.84 x $3.50, or $6.44, as recorded in Column 9.

It will be seen that labor units are not marked on every line. In case an omission is made, the labor unit which embraces this particular material is included with some other item of which it is a part. Thus, switch plates are part of the complete switch outlet, attaching of the plate being included with time for connecting the switch. In the same way, the box hanger is part of the complete outlet box assembly.

ESTIMATING FORM

ITEM	DESCRIPTION	No.	FEET	UNIT COST	MATERIAL COST	UNIT LABOR	HOURS	LABOR COST
1	No.14-2 Type NM Cable		459	4.07C	18.68	.4	1.84	6.44
2	No.14-3 Type NM Cable		184	7.68C	14.13	.4	.74	2.59
3	No.12-2 Type NM Cable		119	5.07C	6.03	.4	.48	1.68
4	4" Octagon Box	7		29.30C	2.05	.2	1.4	4.90
5	4" Square Box	7		43.00C	3.01	.2	1.4	4.90
6	4 1/16" Square Box	3		52.00C	1.56	.2	.6	2.10
7	Switch Recep Box	42		29.20C	12.26	.2	8.4	29.40
8	4" Rd Plaster Ring	7		12.10C	.85	.05	.35	1.23
9	4" Sq Plaster Ring	4		16.60C	.66	.05	.2	.70
10	4" Sq Single Sw Ring	2		22.00C	.44	.05	.1	.35
11	4" Sq Two-Gang Sw Ring	1		22.00C	.22	.05	.05	.18
12	4 1/16" Sq Plaster Ring	2		26.00C	.52	.05	.1	.35
13	4 1/16" Sq Two-Gang Sw Ring	1		28.00C	.28	.05	.05	.18
14	SP Toggle Switch	6		.50	3.00	.2	1.2	4.20
15	3-W Toggle Switch	10		.76	7.60	.3	3.0	10.50
16	4-W Toggle Switch	1		1.45	1.45	.4	.4	1.40
17	Std Plug Recep	19		.42	7.98	.2	3.8	13.30
18	Grd Plug Recep	12		.52	6.24	.2	2.4	8.40
19	Single Sw Plate	13		.09	1.17			
20	Two-Gang Sw Plate	2		.17	.34			
21	Recep Plate	30		.09	2.70			
22	Special Recep Plate (WP)	1		.64	.64			
23	Wirenut	26		1.22C	.32			
24	Connector (1" Box)	6		.13	.78			
25	Box Hanger	13		25.90C	3.37			
26	Ground Strap	3		.30	.90			
27	1" Entrance Cap	1		1.02	1.02	.3	.3	1.05
28	1" Galv Rigid Conduit		10	33.88C	3.39	.9	.09	.32
29	1" Galv Locknut	2		.04	.08			
30	1" Grd Bushing	1		.32	.32			
31	1" Galv Pipe Straps	3		.02	.06			
32	1" Meter Socket	1		3.25	3.25	1.0	1.0	3.50
33	60-A-3P3N Service Sw	1		7.45	7.45	1.9	1.9	6.65
34	1" Cast Ground Clamp	1		.73	.73	1.0	1.0	3.50
35	No.6 Type RW Copper Wire		36	7.97C	2.87	2.0	.72	2.52
36	1" Conduit Nipple	1		.45	.45			
37	Armored Grd Wire		10	16.00C	1.60			
38	Distribution Panel (C-B)	1		15.00	15.00	2.5	2.5	8.75
39	No.8-3 Type NM Cable		34	23.40C	7.96	.5	.5	1.75
40	Connector (3/4" Box)	5		.13	.65			

Fig. 7 Estimating form (*Sheet 1*)

ESTIMATING FORM

ITEM	DESCRIPTION	No.	FEET	UNIT COST	MATERIAL COST	UNIT LABOR	HOURS	LABOR COST
41	40-A CTG	2		.22	.44			
42	Solder (100 Splices)	2 LB		.90	1.80	.07	7.0	24.50
43	Soldering Paste •	1 C		.20	.20			
44	Insulating Tape	2 R		.70	1.40			
45*	No.14 Bare Copper Wire		110	.01	1.10	.05	.6	2.28
46	Hall Chimes	1		6.45	6.45			
47	Rear Chimes	1		4.95	4.95			
48	Transformer	1		2.55	2.55			
49	Front Push Button	1		1.25	1.25			
50	Rear Push Button	1		.30	.30			
51	Bell Wire		160	.01	1.60	1.5	1.5	5.25
52	8-D Nails	1½ LB		.20	.30			
53	1¼" No.10 Wood Screws	12		1.20 C	.10			
54	Type A Ltg Fixture	3		4.15	12.45	.5	1.5	5.25
55	Type B Ltg Fixture	3		2.90	8.70	.4	1.2	4.20
56	Type C Ltg Fixture	1		2.90	2.90	.4	.4	1.40
57	Type D Ltg Fixture	1		2.15	2.15	.4	.4	1.40
58	Type E Ltg Fixture	2		2.85	5.70	.4	.8	2.80
59	Type F Ltg Fixture	1		3.70	3.70	.5	.5	1.75
60	Type G Ltg Fixture	1		8.90	8.90	.7	.7	2.45
61	Type H Ltg Fixture	1		.70	.70	.3	.3	1.05
62	½" EMT (telephone)		20	.80	1.60	4.5	.9	3.15
63	½" EMT Straps	2		.02	.04			
64	Switch Box (Tel)	1		.27	.27	.1	.1	.35
65	½" EMT Connector	1		.12	.12			
66	½" EMT Coupling	1		.10	.10			
67	Sw Plate (Tel)	1		.09	.09			
					211.87			177.17
	Material Cost 211.87							
	Labor Cost 177.17							
	Total Cost 389.04							
	Labor Rate $3.50·Hour							
*	45 - Includes 13 Local Grounds at .05 Hr Each							

Fig. 7. Estimating form (*Sheet 2*)

Time for running the No. 8 armored ground wire is included in the operation of installing service grounds, and the labor for this task is entered opposite *Cast Ground Clamp.* In many cases, there is a choice with respect to labor notation, particularly where the operation is special or "lumped." Thus, the labor for the bell system is entered in line with *Bell Wire.* It could just as well have been entered at *Transformer.*

When all calculations have been performed, Columns 6 and 9 are totalled to find *Material Cost and Labor Cost* for the whole job. Here, cost of material is $211.87, while cost of labor is $177.17. The total cost of material and labor is $389.04.

Final Considerations

The above amount is the estimated prime cost for doing the work. In order to remain in business, it is necessary for the contractor to add a percentage which covers overhead expenses and a margin of profit. The percentage added depends upon a number of factors which can be learned only through experience. A common value is 30 percent. In the present case this added amount is equal to: 0.3 x $389.04, or $116.77. The price which must be quoted to the customer, then, is equal to $389.04 + $116.77, or $506 approximately.

Although the estimate given here is based upon use of non-metallic sheathed cable, the principles involved may be applied to other types of installation, such as knob-and-tube, armored cable, or thin-wall conduit. The take-off operation will be identical, except for minor variation in materials, and some labor units may require adjustment for like reason. The systematic steps outlined in the example will lead to a reasonable estimate in any case.

Short-Cut Method for Estimating. After completing several wiring jobs of the same general nature, a shorter method of estimating can be worked out. The procedure is based upon a knowledge of the average cost of installing individual outlets. This cost, once determined, will be found rather consistent. The cost of a service installation varies according to length of service run and required size of equipment. Lighting fixture costs are more unpredictable because of owner preferences. Bell wiring costs may be readily "lumped." Costs of the telephone outlet varies from job to job. If these variable costs are subtracted from prime cost, and the remainder is divided by the total number of outlets (lighting, switches, and receptacles), a basic outlet cost is found. The above example will again be used.

Prime cost here is $389.04. The service cost is approximately $40, fixtures $65, bell system $18.50, and telephone $6. Subtracting these amounts from prime cost leaves a remainder of about $260, the basic cost for outlets alone. Since there are 62 outlets, the basic cost per outlet is equal to $260 divided by 62, or about $4.20 per outlet. If 30 percent is added to this value, the selling cost per outlet becomes 1.3 x $4.20, or $5.46. In round figures, this amounts to, say, $5.50 per outlet. A similar job may be quickly estimated by multiplying the number of outlets by this value, and adding enough to cover service, fixtures, bell system, and telephone—after weighing these items separately.

As his operations increase in scope, the contractor may lower the unit cost through quantity buying of material and also by cutting down on the average labor charge per hour. The usual method for accomplishing the latter is to take advantage of journeyman and apprentice wage scales. In the present case, if an 80 percent apprentice is employed in company with a $3.50 journeyman, the average labor rate per hour will be lowered. The apprentice will receive: 0.8 x $3.50, or $2.80 per hour. The average rate per hour will be equal to ½ x ($3.50 + $2.80), or $3.15 per hour, instead of $3.50.

REVIEW QUESTIONS

1. Will a very careful estimate provide an exact cost?
2. Are detailed electrical drawings always available to the estimator?
3. Essentially, what kind of a procedure is estimating?
4. What is the first thing that an estimator should do with a plan?
5. Is wire listed in the branch circuit schedule?
6. May the laundry plug receptacle be connected to the laundry lighting circuit?
7. Upon what basis is the branch circuit material schedule constructed?
8. To what point are ceiling outlets measured?
9. To what point are switch outlets measured?
10. How are the short lengths of conductor projecting from each outlet provided for in the total measurement of wire and cable?
11. Was four-wire cable used on the estimated project?
12. When are 4″ square outlet boxes used in place of 4″ octagon boxes?
13. Are labor units expressed in minutes?
14. What labor unit includes the boring of holes in timbers?

15. Is bell wiring labor calculated on an operation by operation basis?

16. How is the distribution panel grounded?

17. What is another name for "local" grounds?

18. In running 500′ of cable, by what value must the labor unit be multiplied in order to determine the number of hours?

19. What is meant by the term *prime cost?*

20. Upon what sort of cost per outlet is the short-cut estimating procedure based?

APPENDIX

The following tables are reprinted from the National Electrical Code by permission of the National Fire Protection Association. The Code is published by the National Board of Fire Underwriters and single copies may be obtained at no charge from their offices below:

85 John St., New York 38, N.Y.
222 West Adams St., Chicago 6, Ill.
465 California St., San Francisco 4, Calif.

Table 220-4(a). Calculation of Feeder Loads by Occupancies

Type of Occupancy	Portion of Lighting Load to which Demand Factor Applies (wattage)	Feeder Demand Factor
Dwellings — other than Hotels	First 3000 or less at Next 3001 to 120,000 at Remainder over 120,000 at	100% 35% 25%
*Hospitals	First 50,000 or less at Remainder over 50,000 at	40% 20%
*Hotels — including Apartment Houses without provision for cooking by tenants	First 20,000 or less at Next 20,001 to 100,000 at Remainder over 100,000 at	50% 40% 30%
Warehouses (Storage)	First 12,500 or less at Remainder over 12,500 at	100% 50%
All Others	Total Wattage	100%

*The demand factors of this Table shall not apply to the computed load of sub-feeders to areas in hospitals and hotels where entire lighting is likely to be used at one time; as in operating rooms, ballrooms, or dining rooms.

TABLE 220-4 (a), NATIONAL ELECTRICAL CODE

Table 220-5. Demand Loads for Household Electric Ranges, Wall-Mounted Ovens, Counter-Mounted Cooking Units and Other Household Cooking Appliances Over 1¾ kw Rating

Column A to be used in all cases except as otherwise permitted in Note 4 below.

NUMBER OF APPLIANCES	Maximum Demand (See Notes)	Demand Factors (See Note 4)	
	COLUMN A (not over 12 kw Rating)	COLUMN B (Less than 3½ kw Rating)	COLUMN C (3½ kw to 8¾ kw Rating)
1	8 kw	80%	80%
2	11 kw	75%	65%
3	14 kw	70%	55%
4	17 kw	66%	50%
5	20 kw	62%	45%
6	21 kw	59%	43%
7	22 kw	56%	40%
8	23 kw	53%	36%
9	24 kw	51%	35%
10	25 kw	49%	34%
11	26 kw	47%	32%
12	27 kw	45%	32%
13	28 kw	43%	32%
14	29 kw	41%	32%
15	30 kw	40%	32%
16	31 kw	39%	28%
17	32 kw	38%	28%
18	33 kw	37%	28%
19	34 kw	36%	28%
20	35 kw	35%	28%
21	36 kw	34%	26%
22	37 kw	33%	26%
23	38 kw	32%	26%
24	39 kw	31%	26%
25	40 kw	30%	26%
26-30	15 kw plus 1 kw for each range	30%	24%
31-40		30%	22%
41-50	25 kw plus ¾ kw for each range	30%	20%
51-60		30%	18%
61 & over		30%	16%

Note 1. Over 12 kw to 21 kw ranges all of same kw rating. For ranges, individually rated more than 12 kw but not more than 21 kw, the maximum demand in Column A shall be increased 5 per cent for each additional kw of rating or major fraction thereof by which the rating of individual ranges exceeds 12 kw.

Note 2. Over 12 kw to 21 kw ranges of unequal ratings. For ranges individually rated more than 12 kw and of different ratings but none exceeding 21 kw an average value of rating shall be calculated by adding together the ratings of all ranges to obtain the total connected load (using 12 kw for any range rated less than 12 kw) and dividing by the total number of ranges; and then the maximum demand in Column A shall be increased 5 per cent for each kw or major fraction thereof by which this average value exceeds 12 kw.

Note 3. Generally, the demand for commercial ranges should be based on the maximum nameplate rating.

Note 4. Over 1¾ kw to 8¾ kw. In lieu of the method provided in Column A, loads rated more than 1¾ kw but not more than 8¾ kw may be considered as the sum of the nameplate ratings of all the loads, multiplied by the demand factors specified in Columns B or C for the given number of loads.

Note 5. Branch Circuit Load. Branch circuit load for one range may be computed in accordance with Table 220-5. The branch circuit load for one wall-mounted oven or one counter-mounted cooking unit shall be the nameplate rating of the appliance.

TABLE 220-5, NATIONAL ELECTRICAL CODE

Table 220-7
Optional Calculation for One-Family Residence

LOAD (in kw or kva)	Per Cent of Load
Air conditioning and cooling including heat pump compressors [see Section 220-4(k)] .....:..............	100%
Central electrical space heating [see Section 220-4(k)]	100%
Less than four separately controlled electrical space heating units [see Section 220-4(k)]	100%
First 10 kw of all other load	100%
Remainder of other load	40%

All other load shall include 1500 watts for each 20 ampere appliance outlet circuit [Section 220-3(b)]; lighting and portable appliances at 3 watts per square foot; all fixed appliances, (including four or more separately controlled space heating units [see Section 220-4(k)], ranges, wall-mounted ovens and counter-mounted cooking units) at nameplate rated load (kva for motors and other low power-factor loads).

TABLE 220-7, NATIONAL ELECTRICAL CODE

**Table 310-12. Allowable Current-Carrying Capacities
of Insulated Copper Conductors in Amperes
Not More than Three Conductors in Raceway or Cable or
Direct Burial (Based on Room Temperature of 30° C. 86° F.)**

Size AWG MCM	Rubber Type R, Type RW, Type RUW (14-2), Type RH-RW See Note 9, Thermoplastic Type T, Type TW	Rubber Type RH, RUH (14-2), Type RH-RW See Note 9, Type RHW, Thermoplastic Type THW	Paper, Thermoplastic Asbestos Type TA, SA, Var-Cam Type V, Asbestos Var-Cam Type AVB, MI Cable, RHH†	Asbestos Var-Cam Type AVA, Type AVL	Impregnated Asbestos Type AI (14-8), Type AIA	Asbestos Type A (14-8), Type AA
14	15	15	25	30	30	30
12	20	20	30	35	40	40
10	30	30	40	45	50	55
8	40	45	50	60	65	70
6	55	65	70	80	85	95
4	70	85	90	105	115	120
3	80	100	105	120	130	145
2	95	115	120	135	145	165
1	110	130	140	160	170	190
0	125	150	155	190	200	225
00	145	175	185	215	230	250
000	165	200	210	245	265	285
0000	195	230	235	275	310	340
250	215	255	270	315	335	...
300	240	285	300	345	380	...
350	260	310	325	390	420	...
400	280	335	360	420	450	...
500	320	380	405	470	500	...
600	355	420	455	525	545	...
700	385	460	490	560	600	...
750	400	475	500	580	620	...
800	410	490	515	600	640	...
900	435	520	555	...	...	...
1000	455	545	585	680	730	...
1250	495	590	645	...	...	...
1500	520	625	700	785	...	...
1750	545	650	735	...	...	...
2000	560	665	775	840	...	...

CORRECTION FACTORS, ROOM TEMPS. OVER 30° C. 86° F.

C.	F.						
40	104	.82	.88	.90	.94	.95	...
45	113	.71	.82	.85	.90	.92	...
50	122	.58	.75	.80	.87	.89	...
55	131	.41	.67	.74	.83	.86	...
60	140	...	.58	.67	.79	.83	.91
70	158	...	.35	.52	.71	.76	.87
75	167	...	...	.43	.66	.72	.86
80	176	...	...	.30	.61	.69	.84
90	194	...	...	...	.50	.61	.80
100	212	...	...	...	...	.51	.77
120	248	...	...	...	...	...	.69
140	284	...	...	...	...	...	.59

†The current-carrying capacities for Type RHH conductors for sizes AWG 14, 12 and 10 shall be the same as designated for Type RH conductors in this Table.

TABLE 310-12, NATIONAL ELECTRICAL CODE

Table 310-13. Allowable Current-Carrying Capacities of Insulated Copper Conductors in Amperes
Single Conductor in Free Air   (Based on Room Temp. of 30° C. 86°F.)

Size AWG MCM	Rubber Type R, Type RW, Type RU (14-2), Type RUW (14-2), Type RH-RW Note 9, Thermoplastic Type T, Type TW	Rubber Type RH, RUH (14-2), Type RH-RW, See Note 9, Type RHW, Thermoplastic Type THW	Thermoplastic Asbestos Type TA, SA, Var-Cam Type V, Asbes Var-Cam Type AVB, MI Cable, RHH†	Asbestos Var-Cam Type AVA, Type AVL	Impregnated Asbestos Type AI (14-8), Type AIA	Asbestos Type A (14-8), Type AA	Slow-burning Type SB
14	20	20	30	40	40	45	30
12	25	25	40	50	50	55	40
10	40	40	55	65	70	75	55
8	55	65	70	85	90	100	70
6	80	95	100	120	125	135	100
4	105	125	135	160	170	180	130
3	120	145	155	180	195	210	150
2	140	170	180	210	225	240	175
1	165	195	210	245	265	280	205
0	195	230	245	285	305	325	235
00	225	265	285	330	355	370	275
000	260	310	330	385	410	430	320
0000	300	360	385	445	475	510	370
250	340	405	425	495	530	....	410
300	375	445	480	555	590	....	460
350	420	505	530	610	655	....	510
400	455	545	575	665	710	....	555
500	515	620	660	765	815	....	630
600	575	690	740	855	910	....	710
700	630	755	815	940	1005	....	780
750	655	785	845	980	1045	....	810
800	680	815	880	1020	1085	....	845
900	730	870	940	....	....	....	905
1000	780	935	1000	1165	1240	....	965
1250	890	1065	1130	....	....	....	....
1500	980	1175	1260	1450	....	....	1215
1750	1070	1280	1370	....	....	....	....
2000	1155	1385	1470	1715	....	....	1405

CORRECTION FACTORS, ROOM TEMPS. OVER 30° C. 86° F.

C. F.							
40 104	.82	.88	.90	.94	.95	...	...
45 113	.71	.82	.85	.90	.92	...	...
50 122	.58	.75	.80	.87	.89	...	...
55 131	.41	.67	.74	.83	.86	...	...
60 140	...	.58	.67	.79	.83	.91	...
70 158	...	.35	.52	.71	.76	.87	...
75 167	...	...	.43	.66	.72	.86	...
80 176	...	...	.30	.61	.69	.84	...
90 194	...	...	...	.50	.61	.80	...
100 212	...	...	...	...	.51	.77	...
120 248	...	...	...	...	...	.69	...
140 284	...	...	...	...	...	.59	...

†The current-carrying capacities for Type RHH conductors for sizes AWG 14, 12 and 10 shall be the same as designated for Type RH conductors in this Table. Weatherproof-covered conductors used for service drops have the same current-carrying capacity as Type SB.

TABLE 310-13, NATIONAL ELECTRICAL CODE

Dictionary of Electrical Terms

A

A.C.: Abbreviation for alternating current.

abscissa: A distance measured horizontally to the right and left of a vertical line.

absolute units: A unit of measurement which has been determined from certain physical properties and upon which all other units are based.

accumulator: A storage battery.

acid proof paint: A paint made especially to resist the action of acid.

admittance: A unit used in alternating current circuits, which is the opposite to impedance, measured in ohms.

aerial: Wires supported above ground and used for receiving or sending electrical waves.

advance wire: An alloy of copper and nickel used for electric heating units.

air blast transformer: A transformer cooled by forcing a circulation of air around its windings.

air gap: Air space between magnetic poles. Space between stationary and rotating parts of an electric motor and generator.

algebraic: Taking into account the sign used in algebra.

alive: Carrying a voltage or current.

alkaline battery: A storage battery using an alkali instead of acid for electrolyte Edison Cells.

all-day-efficiency: The total output divided by the total input of energy for the entire day.

alloy: A metal composed of two or more different metals.

alphaduct: A flexible non-metal conduit.

alternating current: An electric current that reverses its direction of flow at regular intervals.

alternation: One vibration instead of a cycle. One-half a cycle of alternating current.

alternator: An electric generator producing alternating current.

aluminum: A white metal, light in weight but having a higher electrical resistance than copper.

aluminum cell arrester: A lightning arrester using a series of aluminum plates and an electrolyte which forms a thin insulating film on the plates at normal voltage, but becoming a conductor when a high voltage, like lightning, occurs. As soon as the high voltage is reduced to normal, the insulating film is formed again.

aluminum rectifier: A jar containing aluminum plates and iron or lead plates immersed in a solution of ammonium phosphate and which will allow current to flow, though only in one direction, from the iron or lead plates to the aluminum plate.

amalgam: An alloy of mercury or quicksilver with other metals.

amber: A yellowish resinous substance that can be used to produce static electricity by friction.

American wire gauge: The gauge is used for designating the sizes of solid copper wires used in United States. Formerly called Brown and Sharpe (Best gauge).

ammeter: The instrument that indicates the rate of flow of electricity through a circuit.

ammeter shunt: A special low resistance conductor connected to the terminals of an ammeter so as to carry nearly all the current, allowing only a very small current to flow through the instrument itself.

ampere: The practical unit that indicates the rate of flow of electricity through a circuit.

ampere-hour: The quantity of electricity delivered by a current of one ampere flowing for one hour. Used in rating storage batteries.

ampere-hour meter: An instrument that registers or records the number of ampere-hours of electrical energy that have passed througn a circuit.

ampere-turn: The amount of magnetism or magnetizing force produced by a current of one ampere flowing around a coil of one turn. The product of the current flowing through a coil by the number of turns or loops of wire on the coil.

amplifier: A device by which weak currents or sounds acting on another circuit are increased in strength.

anchor: A metal p l a c e d in the ground and to which a guy wire from a pole is attached.

angle of dip: The number of degrees that one end of a magnet dips or points downward.

angle of lag and lead: The distance expressed in degrees that an alternating current lags or leads the voltage wave. The cosine of this angle is called the power factor.

anion: The ion which moves toward the anode in an electrolytic cell.

anneal: To soften by heating and allowing to cool slowly.

annunciator: An electric signal equipment having a number of push buttons located at different places which are wired to an electromagnet in the annunciator box. Press any push button and it causes a signal to be displayed showing what button was operated.

annunciator wire: A soft copper wire that has two layers of cotton threads wound on it in opposite direction and covered with paraffin wax.

anode: The terminal or electrode through which current flows into the electrolyte.

antenna: Wires arranged to receive or send out electromagnetic (Radio) waves into the air.

anti: A prefix, meaning opposite, against, opposed to, etc., to the word that follows it.

apparent E.M.F.: The apparent voltage as measured by the drop in pressure due to current flowing through the resistance.

apparent efficiency: In alternating current apparatus it is the ratio of net power output to volt-amperes input.

apparent watts: The product of volts times amperes in an alternating current circuit.

arc: The flow of electric current across a gap in a circuit which causes a light or glow.

arc furnace: An electric furnace in which heat is produced by an arc between two electrodes.

arc lamp: A lamp producing light from an arc.

arc lamp carbon: A carbon rod between which the arc is produced in an arc lamp.

arc light generator: A generator producing a constant current for an arc light circuit. Nearly obsolete.

arc welding: Joining two pieces of metal together by use of an electric arc.

argon: An odorless, colorless, inert gas taken from the air. Used in some types of incandescent light bulbs.

armature: The rotating part of a direct current motor or generator. The part of the generator that delivers electrical energy or the part of the motor that receives electrical energy from the circuit. Also a piece of iron or steel joining the poles of an electromagnet.

armature air gap: The air space between the stationary and rotating parts of a motor or generator where the magnetic lines of force pass from one to the other.

armature back ampere turns: The magnetic field produced by current flowing in the armature winding that opposes and reduces the number of magnetic lines of force produced by the field magnets of a motor or generator.

armature band: A group of wires wound closely together, or a metal band placed on the coils of the armature to hold them in place.

armature bar: Copper bars used in place of wire winding the armatures of large generators and motors.

armature bore: The space between opposite pole pieces in which the armature revolves.

armature circuit: The path that the current takes in flowing through the windings from one brush to another.

armature coil: The loop or coil of copper wire placed on the armature core and which forms part of the winding.

armature core: The laminated iron part of the armature, formed from thin sheets or disks of steel, and on which the windings are placed.

armature current: The current flowing from the armature of a generator or to the armature of a motor. It does not include the current taken by the shunt field coils.

armature disks: Thin sheets of iron or steel used in building up the armature core.

armature demagnetization: The reduction in the effective magnetic lines of force produced by the armature current.

armature reaction: The effect that the magnetism produced by the current flowing in the armature has on the magnetism (magnetic lines of force) produced by the field coils.

armature resistance: The resistance of the wire used in the windings of the armature measured between rings or brushes, or from positive to negative terminal.

armature slot: The groove or slot in the armature core into which the coils or windings are placed.

armature stand: A device for supporting or holding an armature by the shaft when it is being wound or worked on.

armature tester: Any device or instrument used for locating faults or defects in the armature winding.

armature tooth: The metal between the slots in an armature core.

armature varnish: A liquid put on the field and armature windings to improve the insulation of the cotton coverings on the wires.

armature winding: All of the copper wire placed on the armature and through which current flows when the machine is operating.

armored cable: Rubber - covered wires that have been covered with an iron, steel, or other flexible metallic covering. Often called BX.

artificial magnet: A manufactured permanent magnet, as distinguished from natural magnets.

asbestos: A mineral fiber formed from a certain rock. It is a poor conductor of heat and can withstand high temperatures. Used to insulate wires exposed to a high temperature.

astatic system: An arrangement of two parallel magnets with the north end of one pointing the same way as the south end of the other, so that the two together make a very poor compass needle.

astatic galvanometer: A galvanometer in which the moving parts are arranged in an astatic system or manner.

astatic meter: A meter in which the moving part of element is arranged in an astatic system.

asynchronous: Not having the same frequency; not synchronous; not in step or phase.

asynchronous generator: An induction generator.

asynchronous motor: An induction motor. A motor whose speed is not synchronous with the frequency of the supply line.

atmosphere: The air surrounding the earth. A pressure of 1 atmosphere is 14.7 pounds to the square inch.

atmosphere electricity: Static electricity produced in the sky or between clouds.

atom: The smallest particle or unit of matter that can be chemically united.

atomic weight: The weight of one atom of a chemical element as compared to the weight of an atom of hydrogen.

atonic interrupter: A special interrupter that can be adjusted to operate at a large number of different frequencies.

attachment plugs: A plug that is screwed into a lamp socket, connecting the two wires from an electrical appliance to the circuit.

attenutation: The weakening of an alternating current that flows along a line that has resistance and capacity or leakage.

Aurora Borealis: A light or glow sometimes seen in the northern sky on certain nights.

auto call: A device that sounds a certain code of signals in various places; in a building or factory.

auto-transformer: A transformer in which one winding or coil serves both for the primary and the secondary circuit.

automatic: A device that is operated by certain changes or conditions in an electric circuit and which is not controlled by any person.

automatic telephone: A telephone system where the connection from one party to another is made by means of automatic switches, without the aid of an operator.

automatic time switch: A switch operated at certain times by means of a clock.

automotive: Self propelled vehicles, such as automobiles, trucks, tractors and motorcyles.

automobile battery: The storage battery used in an electric vehicle. The storage battery used for starting and lighting a gasoline automobile.

automobile fuse: A small fuse used to protect the generator and lighting circuits on an automobile.

auxiliary: E x t r a, or something added to the main one.

auxiliary bus: A second bus that may have a different voltage from the main bus and to which a few machines are connected.

auxiliary circuit: Another circuit besides the main circuit; often a control circuit.

auxiliary switch: A switch operated or controlled by the action of another circuit.

B

b: A symbol used for "susceptance" in an alternating current circuit.

B: A symbol for magnetic flux density.

B.B.: Abbreviation for Best Best iron telephone wire.

B-battery: The radio battery that keeps the plate of an electron tube positive in relation to the filament.

B.S.G.: British Standard Gauge.

B. board: One of the switchboards in a large telephone exchange where one subscriber is connected to another.

B.t.u.: British thermal unit; the heat required to raise the temperture of 1 pound of water 1 degree F.

B.W.G.: Birmingham wire gauge. The same as Stubs' Copper wire gauge.

BX: A term often used for flexible armored cable.

B. & S.: Brown & Sharpe wire gauge which is the same as American wire gauge.

babbitt metal: An alloy of lead, tin, copper, zinc, and antimony used for bearings of electrical machines.

back pitch: The distance between the two sides of an armature coil at the back side of the armature, usually expressed in number of slots.

back ampere turns: The ampere turns on the armature that produce magnetism that opposes that produced by the field coils.

bakelite: A moulded insulating material.

balanced load: Arranging the load equally on the two sides of a three-wire system.

balancer coil: An auto transformer used to provide a neutral wire on a 3-wire system.

balancer set: Two direct current generators or motors coupled together and used to keep the voltage the same on each side of a 3-wire system.

ballastic galvanometer: A type of galvanometer used for measuring the quantity of electricity suddenly discharged through it, from usually a condenser, expressed as the angle through which the movable part turns.

bank of lamps: A number of lamps, connected either in series or in parallel, used as a resistance.

bank of transformers: A number of transformers located at one place and connected to the same circuit.

bar magnet: A straight permanent magnet.

bar windings: Windings composed of copper bars or rods instead of wire.

barometer: An instrument for measuring the pressure of the atmosphere.

barrier: A partition, slab, or plate of insulating material placed between blades of switches, wires, or conductor in order to separate or insulate them.

battery: A number of similar units arranged to work together. A number of primary or storage cells connected either in series or in parallel.

battery acid: The liquid used in a storage battery. This is usually sulphuric acid.

battery box: The box holding the cells forming the battery.

battery capacity: The amount of energy that can be obtained from a storage battery—usually expressed in ampere-hours.

battery case: A battery box.

battery charger: A rectifier used for changing alternating current into direct current for charging a battery.

battery connector: A lead covered link or bar used to connect one terminal of a cell to the terminal of the next cell of a battery.

battery discharger: An adjustable resistance used to test the condition of the battery by discharging the battery.

battery hydrometer: A hydrometer used for testing the specific gravity or density of the electrolyte in a storage battery.

battery oven: An oven into which storage batteries are placed and heated in order to soften the compound that seals the cover to the battery cells.

battery paint: A paint, that will resist the action of acids, used to paint the battery boxes or battery rooms.

battery resistance: The internal resistance of a cell or number of cells. The resistance of the plates and electrolyte measured between the external terminals.

battery steamer: An apparatus used for producing steam inside a storage battery case in order to soften the sealing compound.

bayonet socket: A lamp socket that has two lengthwise slots in the sides of socket and at the bottom the slots make a right angle turn. The lamp base has two pins in it that slide in the slots in the socket. The lamp is held in the socket by being given a slight turn when the pins reach the bottom of the slots.

bearing: That part which holds or supports the shaft.

bearing bracket: That part of the machine extending outward from the frame of a machine and which supports or holds the bearings.

bearing loss: Loss of power due to the friction between the shaft and the bearing of a machine.

bearing metal: A special alloy that has the smallest amount of friction between itself and the rotating shaft. Often used as a lining in the bearing.

bell hanger's bit: A long slim wood bit used to drill through the frame of a building when installing door bells.

bell ringing transformer: A small transformer, slipping the voltage down from 110 volts to about 10 volts, used on a door bell.

B-H curve: A curve that shows the relation between the magnetizing force and the number of lines of force per square inch or centimeter produced in different metals.

Bi: A chemical symbol for bismuth.

bichromate cell: A primary cell consisting of carbon and zinc electrodes immersed in a solution of potassium bichromate and sulphuric acid.

binding posts: Terminals used on apparatus or circuits so that other circuits can be quickly attached.

bipolar: Having only two magnetic poles.

Birmingham wire gauge: Wire gauge used for measuring galvanized iron telephone and telegraph wires; also called Stubs' gauge.

block lead: A form of carbon called graphite.

blow-out coil. An electromagnet used for deflecting the arc between two contacts and thus blowing out the arc.

bond: A short high-grade conductor cable or wire used to connect the end of one rail to the next.

booster: A generator connected in series with a circuit in order to increase the voltage of that circuit.

booster converter: A machine that changes the current from alternating to direct or tae opposite, that has a booster built in as part of the machine.

booster transformer: A transformer used to raise the voltage of an alternating current feeder or circuit.

box connector: An attachment used for fastening the ends of cable to a box.

braided wire: A conductor composed of a number of small wires twisted or braided together.

brake shoe: A metallic casting that bears against the wheel in order to stop the wheel from turning.

brake horsepower: The actual power of a machine measured by use of a Prony brake or dynamometer.

branch circuit: That part of the wiring system between the final set of fuses protecting it and the place where the lighting fixtures or drop cords are attached.

branch cutout: The fuse holder for the branch circuit fuse.

brazing: Uniting two metals by a joint composed of a film of brass or alloy that has a higher melting point than solder.

bridge: A Wheatstone bridge.

bridge duplex: A duplex telegraph system dependent for its operation upon the use of a Wheatstone bridge connection in which the telegraph circuit and an artificial line similar to it are the two arms.

bridging set: A telephone set designed to be connected in parallel with other telephones to a telephone line.

British thermal unit: The amount of heat required to raise the temperature of one pound of water 1 degree F.

Brown and Sharpe gauge: The gauge used in United States for copper wires. Same as American wire gauge.

bronze: An alloy of copper and tin.

brush: A conductor that makes connection between the rotating and stationary parts of an electrical machine.

brush discharge: A faint glowing discharge at sharp points from a conductor carrying high voltages. It occurs at a voltage slightly less than that required to cause a spark or arc to jump across the gap.

brush holder: The device used to hold or guide the brushes against a commutator or slip ring.

brush holder cable: A stranded conductor composed of a large number of copper wires, smaller in size than those used on regular stranded cables.

brush holder spring: A spring used to press the brush against the commutator or slip ring.

brush holder stud: An insulated bolt or rod to which the brush holders are fastened.

brush lag: The distance that the brushes on a motor are shifted against rotation in order to overcome the effect of armature reaction.

brush lead: The distance that the brushes on a generator are shifted with rotation in order to overcome armature reaction.

brush pig-tail: A short braided wire fastened to the brush. It conducts the current from the brush holder to the brush.

brush rocker: A support for the brush holders and studs arranged so the location of the brushes can be shifted around the commutator.

brush yoke: Iron framework or support for the brush holders.

bucking: One electrical circuit or action opposing another one.

buckling: Warping or twisting of storage battery plates due to too high a rate of charge or discharge.

bulb: The glass inclosing part of an incandescent lamp that surrounds the filament.

Bunsen cell: A primary cell using zinc and graphite electrodes.

burn out: Damage to electric machine or conductors caused by a heavy flow of current due to short circuit or grounds.

burning rack: A frame for holding storage battery plates when connectors or straps are being fastened to them.

bus: A contraction for bus bar.

bus bar: The main circuit to which all the generators and feeders in a power station can be connected.

bushing: An insulating tube or sleeve protecting a conductor where it passes through a hole in building or apparatus.

butt joint: A splice or connection formed by putting the ends of two conductors together and joining them by welding, brazing, or soldering.

buzzer: A door bell with the hammer and gong removed.

B-X cable: Trade name for armored cable made by General Electric Co. commonly used to refer to armored cable.

C

C: When used with temperature, refers to centigrade thermometer.

C: Capacity of condenser, usually expressed in farads or microfarads.

Cd: Chemical symbol for cadmium.

C.C.W.: Counterclockwise rotation.

Ckt: Circuit.

C.G.S.: Abbreviation for centimeter gram second units—the centimeter being the unit of length, the gram the unit of weight, and the second the unit of time.

C.P.: An abbreviation for constant potential; also for candle-power of a light.

C.W.: Clockwise rotation.

cabinet: Iron box containing fuse, cutouts, and switches.

cable: A conductor composed of a number of wires twisted together.

cable box: A box which protects the connections or splices joining cables of one circuit to another.

cable clamp: A clamp used to fasten cables to their supports.

cable grip: A clamp that grips the cable when it is being pulled into place.

cable rack: A frame for supporting electric cables.

cadmium: A silvery white metal.

cadmium test: A test of the condition of the positive and negative plates of a storage battery.

calibrate: To compare the readings of one meter with those of a standard meter that is accurate.

calido: A nickel-chrome electrical resistance wire.

call-bell: An electric bell that tells a person or operator that he is wanted.

calorie: The amount of heat required to raise the temperature of 1 gram of water 1 degree centigrade.

calorizing: A process of coating a metal with a fine deposit of aluminum similar to galvanizing with zinc.

cambric tape: A cotton tape that has been treated with insulating varnish.

candelabra lamp: A small size lamp that has a smaller size screw base than the standard lamp base but larger than the miniature base.

candle: A unit of light intensity.

candle-power: The amount of light for a source as compared to a standard candle.

canopy: The exterior part of a lighting fixture that fits against the wall or ceilings, thus covering the outlet box.

canopy switch: A switch fastened to the canopy and used to turn on and off the light in the fixture.

caoutchouc: A crude rubber, known as india rubber.

capacity: Ability to hold or carry an electric charge. The unit of capacity is farad or microfarads.

capacity of a condenser:. The quantity of electricity that a condenser can receive or hold.

capacity reactance: The measure of the opposition to the passage of an alternating current through a condenser expressed in ohms.

capillary attraction: The course of the raising and lowering of the liquid in a tube above or below the surrounding liquid.

carbon: A non-metallic element or substance found in graphite, charcoal, coal, and coke.

carbon brush: A block of carbon used to carry the current from the stationary to the rotating part of a machine.

carbon contact: A contact made of carbon used where the circuit is opened frequently.

carbon disk: A piece of carbon used as a resistance in a rheostat.

carbon holder: A device for holding and feeding the carbon rods in an arc light.

carbon pile regulator: A number of pieces of carbon arranged as a rheostat to regulate the current to another circuit.

carbon resistance: A resistance formed by carbon plates or powder and arranged so that the pressure on the plates can be varied. The less the pressure, the greater will be the resistance.

carbonize: To turn some other material to carbon by fire.

carrier current: A very high frequency current used to provide the energy for transmitting a radio message.

carrying capacity: The amount of current a wire can carry without overheating.

cartridge fuse: A fuse inclosed in an insulating tube in order to confine the arc or vapor when the fuse blows.

case-hardening: The hardening of the outside of metals with heat.

cascade connection: An electrical connection in which the winding of one machine is connected to a different winding of the next machine.

cascade converter: A rotary converter that receives its energy from the rotor (secondary) of an induction motor connected to the same shaft.

cat whisker: A fine wire spring, one end of which makes contact with a crystal in a crystal radio set.

catenary curve: The curve or sag formed by the weight of a wire hanging freely between two points.

cathion: That part of the electrolyte that tends to be liberated at the terminal when the current leaves the electrolyte.

cathode: The electrode toward which the current flows in an electrolyte. The negative electrode or terminal.

cathode rays: Those rays coming from the cathode of a vacuum tube which produce X-Rays when they strike a solid substance in the tube.

cauterize: The searing or burning of flesh with an electrical heated wire.

C.C.: An abbreviation for cubic centimeter; also **Cu c.n.** is used.

cell: A jar or container holding the plates and electrolyte of one unit of a storage or primary battery.

cell vent: An opening in the cover of a cell which allows the gasses found in the cell to escape.

celluloid: An insulating material made from gun cotton and camphor; it ignites easily and burns up very quickly.

cement: A material used to bind substances together.

cementation: The forming of lead sulphate in small quantities on storage battery plates when they are drying after being made.

center of distribution: A point near the center of the area or section served by a feeder or circuit from a power station or substation. The feeder is usually run directly to this point, and then branches out in all directions from there.

centigrade: A thermometer whose scale is 0 at the freezing point and 100 at the boiling point of water.

centimeter: The one-hundredth part of a meter; 0.3937 inches, or longer than ⅜ of an inch.

central: A telephone office or exchange.

central station: A power plant supplying electric light and power to a number of users.

centrifugal cutout: A switch opened by centrifugal force of a rotating body and closed by a spring when the centrifugal force is reduced.

centrifugal force: The force that tends to throw a rotating body, or weight, outward and away from the center of rotation.

chain winding: A type of armature winding which resembles a chain.

characteristic: A curve that shows the ability of a machine to produce certain results under a certain given condition.

charge: That quantity of static electricity stored between the plates of a condenser.

charging: Sending electric current through a storage battery.

charging rate: The number of amperes of current flowing through a storage battery when it is being charged.

choke coil: A coil of a low ohmic resistance and a high inductance which will hold back unusual currents but allow regular steady currents to flow through easily; also reactors or reactance coils.

circuit: The path taken by an electrical current in flowing through a conductor from one terminal of the source of supply to the other.

circuit breaker: A device used to open a circuit automatically.

circular loom: A flexible non-metallic tubing slipped over rubber covered wires for additional insulation and protection.

circular mil: The area of a circle one-thousandth of an inch in diameter; area in circular mils = diameter, in mils, squared or multiplied by itself.

Clark cell: A primary cell that produces a constant voltage for several years and used as a standard source of voltage.

cleat: Piece of insulating material used for fastening wires to flat surfaces.

climber: A sharp steel spur or spike fastened to the shoe and legs of linemen to aid them in climbing poles.

lockwise rotation: Turning in the same direction as the hands of a clock; right-handed rotation.

closed circuit: A complete electric circuit through which current will flow when voltage is applied.

closed circuit battery: Primary cells that will deliver a steady current for a long time. A battery that can be used on a closed circuit system.

closed coil armature: The usual armature windings in which the connection of all coils forms a complete or closed circuit.

closed magnetic circuit: A complete magnetic path through iron or other metal without an air gap.

cluster: A lighting fixture having two or more lamps on it.

cobalt: A white metal similar to nickel.

code: A series of long and short sounds given in order to convey certain signals or information.

coefficient of expansion: The increase in length of a rod or body for each degree that the temperature is increased.

coherer: A device used in the early days of radio to detect radio signals.

coil box: A box containing ignition or induction coils.

coil pitch: The number of slots spanned by an armature coil.

collector ring: A metal ring fastened to the rotating part of a machine, and completing the circuit to the rotating part of the machine.

combination fixture: A fixture arranged for both gas and electric lights.

combination switch: A switch on automobiles used to control both lights and ignition.

commercial efficiency: The ratio of total output to input of power.

commutating machines: Generator, motors, and rotary converters that have commutators.

commutating pole: An interpole placed between the pole pieces of a dynamo in order to reduce sparking at the brushes.

commutating pole rectifier: A rotary converter fitted with interpoles.

commutation: Changing the alternating current produced in the armature windings into direct current by use of the commutator and brushes.

commutator: A device by which alternating current produced in a generator is changed into direct current. It consists of a ring made up of a number of copper bars or segments; each bar is insulated from the next one and connected to the end of the armature windings.

commutator bar: A small piece of copper used in building a commutator; a commutator segment.

commutator cement: An insulating substance used in repairing or replacing mica in a commutator.

commutator compound: A compound applied to the surface of a commutator to assist in obtaining a smooth polish.

compass: A small magnetized needle pivoted at the center and pointing in a north and south direction, which is in line with the earth's magnetism, unless influenced by stronger magnets.

compensated machine: A motor or generator with a series field winding placed in slots in the face of the pole piece.

compensated voltmeter: A voltmeter connected with the bus bars at a power station. It indicates the voltage in the feeders, showing the actual pressure furnished at the far end of the circuit.

compensated winding: A winding which is placed in slots cut in the face of the pole pieces parallel with the armature slots. The current in this winding flows in the opposite direction to that in the armature slots.

compensator: A name that is applied to any device which offsets or equalizes in its effect some undesired effect.

component: A part of any thing; used in reference to the analyzing of a current in a circuit by vectors.

composite line: A telephone or telegraph line composed partly of underground and partly of overhead open wires. A line that telegraph and telephone messages may be sent over at the same time.

compound field winding: A winding composed of shunt and series coils either acting together or against each other.

compound generator: A generator that has shunt and series field coils acting together to produce a steady voltage.

compound magnet: A permanent magnet built up from a number of thin magnets of the same shape.

compound motor: A motor that has shunt and series field coils or windings.

concentrated acid: Pure acid that must be diluted before it can be used.

concentric cable: A number of wires wound spirally around and insulated from a central conductor or cable.

condenser: Two conductors separated by an insulating material that is capable of holding an electrical charge.

condenser capacity: The amount of electrical charge that a condenser will hold, measured in microfarads.

condenser dielectric: Insulating material between condenser plates or conductors.

condenser plate: One of the conductors forming the condenser.

condensite: A kind of moulded insulation.

conductance: The ease with which a conductor carries an electric current; it is the opposite of resistance. The unit of conductance is the mho (word "ohm" spelled backward).

conductivity: The ability of a substance to carry an electric current.

conductor: A wire or path through which a current of elecricity flows; that which carries a current of electricity.

conduit: A pipe or tube, made of metal or other material, in which electrical conductors or wires are placed.

conduit box: An iron or steel box located between the ends of the conduit where the wires or cables are spliced.

conduit bushing: A short threaded sleeve fastened to the end of the conduit inside the outlet box. Inside of sleeve is rounded out on one end to prevent injury to the wires.

conduit coupling: A short metal tube threaded on the inside and used to fasten two pieces of conduit end to end.

conduit elbow: A short piece of conduit bent to an angle, usually to 45 or 90 degrees.

conduit rigid: A mild steel tubing used to inclose electric light and power wires.

conduit rod: A short rod which is coupled to other rods and pushed through the large conduit to remove obstructions and pull a cable into the conduit.

conduit wiring: Electric light wires placed inside conduit.

condulet: The trade name for a number of conduit fittings made by Crouse-Hinds Co.

connected load: The sum of the rating of all the lamps, motors, heating devices, etc., connected to that circuit.

connecting-up: The process in making splices and connections to complete an electric circuit.

connector: A device used to connect or join one circuit or terminal to another.

connector switch: A device in an automatic telephone exchange that makes connection with the desired line.

Consequent pole: A magnetic pole produced by placing together or near each other two north or two south poles. The forming of a pole along a magnet as well as at the ends.

consonant: A condition in a transformer which produces resonance in the primary circuit due to a certain combination of capacity and reactance in the secondary circuit. A condition to be avoided except in radio work.

constant current: A current whose amperage is the same all the time.

constant-current circuit: A series circuit, such as a street lighting circuit.

constant-current generator. A generator in which the voltage is increased as the load increases while the current is kept constant.

constant-current motor: A motor designed to operate on a constant-current circuit.

constant - current transformer: A transformer whose secondary delivers a constant alternating current, usually to a series street lighting circuit. The primary is connected to a constant-potential circuit.

constant potential: A constant voltage or pressure in the usual power and light circuit.

constant-potential generator: A generator that produces a constant voltage even though the speed is varying or changing.

constant- potential transformer: A transformer used on a constant-potential circuit.

constant-speed motor: A motor that runs at the same speed when carrying a full load as when lightly loaded.

constant-voltage regulator: A regulator that causes a generator to produce a steady voltage at varying loads.

contact: A place where a circuit is completed by a metallic point being pressed against a conductor. When the pressure is removed, the circuit is opened and flow of current stopped.

contact drop: The voltage drop across the terminals of a contact.

contact resistance: The resistance in ohms across the contact points.

contact sparking: The spark or arc formed at the contact points when a circuit carrying current is opened.

contactor: A device used to open and close an electrical circuit rapidly and often.

continental code: A series of dot and dash signals generally used in radio work to send telegraph messages.

continuous current: A direct current that is free from pulsations.

continuous rating: The output at which a machine can operate continuously without overheating or exceeding a certain temperature.

contractor: One who agrees to do a certain job for a sum of money agreed upon before the work is started.

control switch: A small switch used to open and close a circuit which operates a motor or an electromagnet coil. This motor or electromagnet is used to operate or control some electric machine.

controller: A device that governs or controls the action of electrical machines connected to it.

controller resistance: The resistance used with a controller to start and vary the speed of the motor.

converter: A machine that changes electric current of one kind into current of another kind by the use of rotating parts.

conveyors: Mechanical devices used to carry material from one place to another.

Coolidge tube: An X-ray tube first developed by Wm. D. Coolidge.

copper: A metal used for electrical conductors because it has less resistance than any other metal except silver.

copper bath: An electrolyte composed of copper salts or crystals used for copper plating.

copper clad: Iron or steel wire covered with a layer of copper in order to increase the conductivity

copper loss: The I^2R loss in power due to the resistance of the copper conductors or wires.

copper plating: Depositing a layer of copper or other metals by the electroplating process.

copper ribbon: A thin bar or strip of copper.

copper strip: A long thin bar of copper, usually about $\frac{1}{8}$ to $\frac{3}{8}$ of an inch thick.

cord: Two insulated flexible wires or cables twisted or held together with a covering of rubber, tape, or braid.

core: The iron or steel in the center of a coil through which magnetic lines of force pass.

core iron: Iron sheets used for making cores of magnets, transformers, generators, and motors.

core loss: The power lost in a machine due to eddy currents and hysteresis losses.

core transformer: A transformer with the windings placed on the outside of the core.

corona: A violet light glow that occurs on high voltage conductors just before the voltage becomes high enough to cause a spark or arc.

corrosion: The rusting of iron and a similar action and deposit formed on other metals.

cotton-covered wire: A wire covered with a layer of thin cotton threads wound spirally around it.

cotton-enameled wire: An enameled insulated wire covered with a layer of cotton threads.

cotton sleeving: A woven cotton sleeve or tube slipped over wires to insulate them.

coulomb: The quantity of electricity passing through a circuit. It is equal to amperes times seconds. An ampere hour = 3600 coulombs.

counter-clockwise rotation: Turning left handed, which is in a direction opposite to that of the hands of a clock.

counter-electromotive force: The voltage or pressure that opposes the normal voltage tending to force a current through a circuit.

cowl lamp: A lamp placed on the dashboard of an automobile to light the instruments on it.

creeping of wattmeter: A slow turning of the wattmeter disk when there is no power passing through it.

Crookes' tubes: Tubes used for producing X-rays.

cross arm: An arm fastened at the top of the pole to support the wires.

cross magnetization: The magnetic lines of force produced in the armature that are at right angles to those produced by field coils.

cross over: A device that enables one wire to cross over another or a car to pass from one track to another parallel one.

cross-section area: The surface of the end of a wire, rod, or other object. It is measured in square inches, square centimeters, square mills, etc.

crow foot: A small fitting fastened in an outlet box to which fixtures are fastened.

crow-foot zinc: A zinc plate having extending arms, used in a gravity cell.

current: The flow of electricity through a circuit.

current coil: The coil or winding through which the current in a circuit flows.

current density: The number of amperes per square centimeter or square inch of cross sectional area of the conductor.

current regulator: A device that regulates or limits the flow of current through a circuit.

current strength: The flow of current in amperes.

current transformer: A transformer in which the flow of current in the secondary winding is in proportion to that flowing through the primary circuit, also called series transformer.

cut-in: A device operating in an electric circuit which connects two circuits together.

cut-out: A device that opens or disconnects one circuit from another.

cut-out box: The box in which fuse holder blocks, and fuses are located.

cycle: The flow of alternating current first in one direction and then in the opposite direction in one cycle. This occurs 60 times every second in a 60-cycle circuit.

D

D.C.: Used as an abbreviation for "direct current." Used as an abbreviation for "double contact."

D.C.C.: Used as an abbreviation for "double cotton-covered wire."

D.P.: Used as an abbreviation for "double pole."

D.P.S.: Used as an abbreviation for "double pole snap switch."

D.P.S.T.: Used as an abbreviation for "double pole single throw."

D.P.D.T.: Used as an abbreviation for "double pole double throw."

damper winding: As applied to copper pieces so placed in the pole faces of alternating-current machines as to reduce hunting.

damping: Causes the needle of an electric measuring instrument to come to rest quickly.

damping coil: Used to cause the needle of a galvanometer to quickly return to zero.

damping magnet: Any magnet used to check the motions of a moving object or magnet.

Daniel cell: A primary electric cell, using copper and zinc for electrodes, used on closed circuit work.

D'Arsonval meter: A voltmeter or ammeter whose pointer is attached to a moving coil of fine wire carried between the poles of a permanent magnet.

dashboard instruments: Ammeter, voltmeter, or current indicator, suitable for mounting on the dashboard or cowl board of an automobile.

dash pot: A cylindrical chamber containing oil, air, or other fluid in which moves a plunger attached to some part in which it is desired to avoid sudden changes of position.

dead beat: An instrument whose pointer comes immediately to its true reading without swinging back and forth.

dead coil: An armature coil which is not connected in the armature circuit of the windings but which is required in order that there may be the proper number of coil sides in each slot.

dead end: The end of a wire to which no electrical connection is made. The end used for supporting the wire. The part of a coil or winding that is not in use.

dead end eye: A metal eye threaded at one end to attach to a rod and holding a cable in the loop of the

dead ground: An accidental ground of low resistance through which most of the current can escape from a circuit.

dead man: A short pole with cross-arms to which the guy wire from another pole is fastened.

dead wire: A wire in which there is no electric current or voltage.

decade bridge: A Wheatstone bridge having ten separate coils of equal resistance value.

deci: Is a term meaning one-tenth.

deci-ampere: One-tenth of an ampere.

declination: The difference between the position of a compass needle and the true position of geographical north and south.

declinometer: An instrument for measuring the declination of a compass needle.

de-energize: To stop current from flowing in a circuit or an electrical part.

deflection: The movement of the indicating pointer of an electric measuring instrument.

deflection of compass needle: The movement of a needle from a point of repose either in the earth's magnetic field or in that of another magnet and produced by the influence of the flux of an electric current or of a magnet.

deka: A prefix meaning ten times.

deka ampere: Ten amperes.

delivered power: The power delivered at one end of a line, in a system of electrical transmission in contradistinction to the power delivered into the line at the other end.

delta connection: Series hookup of three circuits of an alternator, the end of one circuit being connected to the beginning of the next, etc. The wiring diagram of this arrangement resembles a triangle or the letter Delta of the Greek alphabet.

demagnetization: Process of removing the magnetism from a magnetized substance. This may be done either by heating to a red heat, by violent jarring, or by holding the magnetized substance in and then gradually removing it from the magnetic field of a solenoid operated on an alternating current.

demagnetizing armature turns: Inductors of an armature, which, while moving in the field of the poles, set up a counter-magnetic field that tends to demagnetize the poles.

demand: Amount of electric current needed from a circuit or generator.

demand factor: Ratio of the maximum amount of current consumed in one sub-circuit to the total load or current draw on the whole circuit.

demand meter: Device which registers the maximum ampere consumption of appreciable duration in a circuit.

density: The ratio of a quantity of a substance to the space it occupies; i.e., the ratio of mass to volume.

density of current: Amount of current flowing through a conductor of given cross-sectional area.

density of field: Amount of magnetic flux, or lines of force, contained in a given cross-sectional area.

density of electrolyte: The proportion of chemical in the water with which it is mixed to make an electrolyte. See "specific gravity."

depolarize: (a) To eliminate or retard the gas which tends to collect on the electrodes of an electric cell when it is being charged or discharged. (b) Synonym for demagnetize.

depolarizer: A chemical, electrochemical, or mechanical agent introduced into the cell to prevent or retard the formation of gas which polarizes the electrodes.

derived circuit: Shunt or parallel circuit, the current for which is obtained from another circuit.

deviation factor: Difference between an alternating-current wave of a generator and a true sine wave.

diamagnetic substance: One that is repelled by a magnet, as bismuth and phosphorus.

diaphragm: A disk or sheet of metal or other substance having enough flexibility to vibrate, as a telephone-receiver diaphragm.

dielectric: Insulation between conductors of opposite polarity; term generally used only when induction may take place through it.

dielectric constant: A number representing the dielectric quality of a given substance as compared to that of air.

dielectric current: Leakage of current through a dielectric.

dielectric hysteresis: Consumption of energy caused by molecular friction in a dielectric under changes of electrostatic pressure.

dielectric resistance: Resistance of a dielectric to electrical pressure.

dielectric strain: Strain to which a dielectric is subjected while it is under electrical pressure.

dielectric strength: Ability of a dielectric to withstand electrical pressure before breaking down. This is measured in volts necessary to puncture the dielectric.

dies (pipe): Tools for cutting and threading metal conduit.

difference of potential: Difference in voltage between two conductors or two points along one conductor carrying an electric current.

differential booster: Generator in a battery-charging arrangement to maintain a constant voltage.

differential electromagnet: An electromagnet having part of its winding reversed to oppose the other part to permit adjustment of the pull.

differential field winding: Field winding in which the shunt and series windings of a compound-wound motor or generator oppose each other.

differential galvanometer: Galvanometer having two coils wound to counteract each other.

differential generator: Generator in which the shunt and series field windings counteract each other to limit the maximum amperage.

differential motor: A direct-current motor having its shunt and series field windings opposing each other, to obtain a constant speed.

differential relay: A relay consisting of a differential electromagnet.

differential winding: Coil which is wound opposite to another to counteract it.

diffusion of magnetic flux: Deviation of the magnetic lines of force from a straight path between the poles.

dimmer: A resistance coil connected in series with a lamp to reduce the amount of current flowing through it, and consequently to dim or reduce the light.

dinkey: Small, two-wheeled cart used for hauling poles in line construction.

dip: Angle which a magnetic needle, pivoted in a vertical plane, makes with the horizontal.

dipping needle: Magnetized needle pivoted freely at its center of gravity in a vertical plane so that, when set in a magnetic meridian, it dips until it lies parallel to the magnetic lines of force of the earth.

diphase generator: Generator producing two alternating currents a quarter of a cycle apart.

diplex telegraphy: Transmission of two telegraphic messages over the same wire, at the same time, and in the same direction.

direct-connected: Two electrical machines, such as a motor and a generator, connected together mechanically and in line, by having their shafts coupled together or by both being mounted on the same shaft.

direct current: Electric current flowing over a conductor in one direction only. Abbreviation d.c.

direct-current convertor: Device for changing a direct current of one potential to a direct current of another potential.

direct-current generator: Generator that delivers direct current.

direct-current instrument: Device operated on direct current.

direct-current magnet: Electromagnet operated on direct current.

direct-reading galvanometer: A galvanometer provided with a scale so calibrated that the current flow may be read directly, without the necessity of calculating it from the proportions of the coil and the magnetic moment of the needle.

disc armature: Armature of a generator consisting of a flat disc on which the coils are mounted.

discharge: Removal of electricity from its source through a circuit.

discharge recorder: Device which detects and records discharges through a lightning arrestor.

discharge resistance: Resistance coil which is connected across a circuit breaker to prevent arcing when the contacts separate.

discharger: Resistance device which is connected across the terminals of a storage battery to discharge it slowly without damaging it.

disconnect: To remove an electrical device from a circuit, or to unfasten a wire, making part or all of the circuit inoperative. The word is particularly applied to the act of severing a telephone connection to permit repairs.

disconnector: Switch for cutting out circuits having high voltages, done only under a minimum load.

displacement current: Small current of electricity in a dielectric which is under strain of a high potential.

disruptive discharge: Violent discharge of electricity accompanied by a spark.

dissonance: Lack of consonance or agreement; as of alternating currents of opposite phase.

dissociation: Separation of the component elements of a chemical mixture or compound, without the aid of any other chemical agency.

distortion of field: A condition causing magnetic flux between the poles of a magnet to assume an arched or curved path instead of a straight one from pole to pole.

distribution box: Small metal box in a conduit installation, giving accessibility for connecting branch circuits.

distributing frame: Structure where connections are made between the inside and outside wires of a telephone exchange.

distribution: Division of current between the branches of an electrical circuit.

distribution lines: The main feed line of a circuit to which branch circuits are connected.

distribution center: Point along the main feed lines which is approximately in the center of the branch lines.

distribution panel: Insulated board from which connections are made between the main feed lines and branch lines.

distribution system : The whole circuit and all of its branches which supply electricity to consumers.

distributive: Tending or serving to distribute.

divided circuits: Approximate division of a distribution system to balance both sides of the lines. A divided magnetic circuit is one having more than one path through which the flux passes.

dome lamp: Small lamp attached to the underside of the top of an automobile.

door lamp: Small lamp for lighting the doorway and running board of an automobile.

door lantern: Lamp hung so as to illuminate the entrance of a house.

door opener: Motor-driven device for opening and closing garage doors.

door switch: Switch which is operated by opening and closing the door to which it is connected.

double armature: An armature which has two separate windings on one core.

double-break switch: Switch which connects and disconnects two contacts at the same time.

double-contact lamp: Lamp with a base having two terminals to which electrical contact is made when it is inserted in a socket.

double-cotton-covered: Wire covered with two layers of cotton insulation. Abbreviation d.c.c.

double-current generator: A generator delivering both direct and alternating current.

double deck: Arrangement of two electrical machines, one mounted above the other.

double delta connection: Connection of three transformers by which a 3-phase system is connected to a 6-phase system.

double-filament lamp: Lamp having two separate filaments of different resistances to provide low and high brilliancy.

double-pole: A term designating two contacts or connections on a device, for instance, a double-pole knife switch. Abbreviation d.p.

double reduction: Speed reduction in a machine obtained by using two sets of gears or pulleys.

double-silk-covered. Wire covered with two layers of silk insulation. Abbreviation d.s.c.

double-throw switch: Switch which can be operated by making contact with two circuits. Abbreviation d.t.

double trolley: Street-railway system using two overhead trolleys instead of one, carrying the positive and negative current. This arrangement eliminates electrolysis caused by grounding one conductor, but it increases trolley troubles, and for this reason is seldom used.

double re-entrant winding: Armature winding, half the conductors of which make a closed circuit.

draft: Air drawn or forced up into the fire-box to accelerate fuel combustion.

draft tube: Tube or passage through which the discharge from a hydraulic turbine flows into the tailrace.

draw bar: Bar on a locomotive which is used to connect it to a train.

draw-bar pull: Force available at the draw bar of a locomotive to pull a train, as distinguished from the actual power of the engine or motor.

drive shaft: Shaft employed to drive a number of machines. Line shaft.

driven pulley: A pulley to which movement is imparted by means of a belt from another pulley.

driving pulley: Pulley that drives another through the medium of a belt.

drop: Usual term for drop of potential.

drop annunciator: Annunciator having one or more electromagnets, each of which, when operated, releases a catch holding a small plate or shutter, and allows it to drop, exposing a number or letter.

drop of potential or voltage: Decrease of voltage at points along a circuit, caused by resistance.

drop wire: Wire which is connected to a feed wire outside of a building, and brings the supply inside.

drum: The laminated iron cylinder or core of an armature for a generator or motor.

dry battery: A number of dry cells connected together in series or parallel to obtain more voltage or amperage, respectively.

dry cell: A primary source of electric current consisting of three elements; a zinc cylinder, a paste electrolyte and a carbon rod or electrode. The zinc cylinder is filled with the electrolyte and the carbon electrode is placed in the center but not touching the zinc; the top of the cell is sealed with a wax compound. The chemical action of the electrolyte on the zinc sets up an electric current when the cell is connected to a current-consuming device, as a bell. The carbon electrode is the positive and the zinc is the negative.

dry storage: Method of keeping a storage battery, when not in use, by removing the electrolyte.

dual ignition: Ignition system for an internal combustion engine which may obtain current from either a battery or a magneto as desired.

dual magneto: Ignition magneto which has its armature wound so that it can deliver both its own current and that of a battery to the distributor.

duct: (a) A space in an underground conduit to hold a cable or conductor. (b) A ventilating passage for cooling an electrical machine.

duct-foot: Unit expressing the total length of all the cableways in one lineal foot of an underground conduit. Thus, a 6-duct conduit, one foot long, contains 6 duct feet.

duo-lateral coils: Form of honeycomb inductance coils used in radio, which are designed to reduce the distributed capacity.

duplex cable: Cable consisting of two wires insulated from each other and having a common insulation covering both.

duplex ignition: Ignition system capable of sending both the battery and the magneto current into the induction coil at the same time.

duplex telegraphy: Telegraph circuit permitting the transmission of two messages in opposite directions at the same time over a single wire.

duplex winding: Two separate windings on the same armature or coil.

duplex wire: Same as duplex cable.

dynamic braking: Method of stopping a motor quickly without the aid of a mechanical brake. On d.c., a resistor connected across the armature changes the electrical energy produced by the motor, which acts as a generator when the line circuit is broken, into heat. On polyphase a.c. motors this method of braking is obtained by energizing one phase winding with direct current.

dynamic electricity: Electricity in motion as distinguished from static electricity.

dynamo: A synonym for generator; formerly applied to both motor and generator, although modern use tends to confine its meaning to d.c. generators.

dynamometer: Mechanical or electrical device for measuring the torque of a machine in order to determine its power output.

dynamotor: Electrical machine which acts as both motor and generator, running on and producing either direct or alternating current. It has one field and two separate armatures or a double-wound armature.

dyne: Unit of force. Power or force required to cause an acceleration of one centimeter per second to a mass of one gram.

E

E: Symbol for volts.

EBB: Abbreviation for "extra best best" iron wire used for telephone and telegraphic purposes.

E.C.&M.: Trade name for electric-control equipment, lifting magnets, etc.

E.H.P.: Abbreviation for electrical horsepower.

E.M.F.: Abbreviation for electromotive force.

E.P.C.: Abbreviation for Electric Power Club.

ear: Device for supporting a trolley line; a bronze casting grooved to receive the trolley and having lips which are clinched around it. The ear is supported on a trolley hanger.

earth: Synonym of "ground," meaning the grounded side of an electrical circuit or machine.

earth current: Current passing through the ground.

ebonite: Substance consisting of black hard rubber and sulphur. It is hard and brittle, has high insulating qualities, and possesses inductive qualities to a high degree.

economizer: Device used on boilers to absorb the heat that has passed the flues and would be wasted out of the stack; used to preheat boiler feed water.

eddy-current loss: Loss of energy of an electrical machine which is caused by eddy currents.

eddy currents: Currents in armatures, pole pieces, and magnetic cores, induced by changing electromotive force. It is wasted energy and creates heat.

Edison battery: Storage battery having plates made of nickel peroxide and iron, and using potassium hydrate and water for an electrolyte.

Edison distributing box: Box used in three-way distribution systems.

Edison-Lalande cell: Primary electric cell having electrodes made of copper oxide and zinc and using a caustic soda solution as an electrolyte.

"Ediswan" or bayonet socket: Lamp socket having a bayonet base, which is popular on automobile-lighting systems, and is used for house lighting in England.

"Ediswan" connector: Plug connector having a base similar to that of an "Ediswan" socket.

effective current: Value of a current as shown on a steady-reading ammeter.

effective electromotive force: Difference between the impressed and the counter e.m.f.

effective resistance: All electrical and inductive losses of a current.

efficiency: The ratio of the amount of power or work obtained from a machine and the amount of power used to operate it.

elastance: Inability or opposition to retaining an electrostatic charge. Opposite to capacity.

elastivity: Specific elastance of a substance.

elbow: Hollow fixture for connecting two lengths of conduit at an angle, usually fitted with a removable cap to facilitate drawing the wires through one conduit and then inserting them in the other one.

Electragist: Term used by the National Association of Electrical Contractors and Dealers, now the Association of Electragists, to denote a person conducting an electrical-contracting business.

electric or electrical: Pertaining to electricity.

electric circuit: Path through which an electric current flows.

electric breeze or wind: Emission of negative electricity from a sharp point of a conductor carrying a high potential.

electric candle: Small electric arc lamp.

electric charge: Quantity of electricity on a conductor.

electric eel: An eel found in South American waters which is capable of giving off painful and dangerous shocks of high potential, estimated equal to the combined charge of 15 Leyden jars, each having 1 2/3 square feet of tinfoil coating.

electric energy: Power of electricity to perform work, mechanically or in the production of heat and light.

electric furnace: Furnace using electricity to produce heat.

electric glow: electrostatic discharge causing a violet light around conductors carrying high potentials, occurring just before the emission of a spark or a steady brush discharge.

electric heater: Heater consisting of resistance wire which becomes hot as the current flows through it.

electric horsepower: The equivalent of one horsepower in electrical energy, which is 746 watts.

electric potential: Pressure or voltage of electricity.

electric power plant: Installation consisting of a prime mover driving a generator to produce electricity.

electric spectrum: The component colors of an electric arc separated by means of a glass prism.

electric units: Standards of measurement of electrical properties; for instance, ampere, volt, ohm, farad, henry, etc.

electric wave: Theoretical form of movement of an electric current transmitted through air.

electric welding: Process of welding with the use of an electric arc or with heat generated by current flowing through the resistance of the work to be welded.

electrical codes: Rules and regulations for the installation and operation of electrical devices and currents.

electrical series: A list of substances which, when two are rubbed together, will produce an electrostatic charge, as silk and hard rubber.

electrical sheet: Steel or iron sheets from which laminations for electrical machines are punched.

Electrician: Person working or experimenting with electrical devices.

electricity: Invisible energy capable of moving 186,000 miles per second. Electricity is really not capable of being defined exactly, with present knowledge.

electrification: (a) Providing means to operate devices with electricity. (b) To impose a static charge.

Electrochemical: Pertaining to the interaction of certain chemicals and electricity, the production of electricity by chemical changes, the effect of electricity upon chemicals, etc.

electrochemistry: Science of electrochemical interaction.

electrodynamic: Pertaining to electricity in action.

electrodynamometer: D e v i c e for measuring the strength of an electric current by its attraction or repulsion to conductors carrying current.

electrocute: (a) To execute a criminal by electricity. (b) Persons accidentally killed by electricity are said to be electrocuted.

electrode: Either terminal of an electric source, particularly an electric cell. Also applied to the terminals of electrical apparatus applied to the human body in the treatment of disease.

electrokinetic: Pertaining to electricity in action.

electrolier: Hanging electric fixture holding lamps which can be lighted separately or all at once.

electrolier switch: Switch which controls the lamps of an electrolier.

electrolysis: Chemical decomposition caused by an electric current.

electrolyte: Chemical solution used in an electrical device which passes an electric current.

electrolytic: Pertaining to electrolysis.

electrolytic condenser: Condenser using an electrolyte as a dielectric.

electrolytic decomposition: Separation of the elements in an electrolyte.

electrolytic generator: Generator for charging storage batteries.

electrolytic interrupter: Device for rapidly interrupting or breaking up a direct current into pulsations, consisting of a cathode, generally a lead plate, immersed in a dilute solution of sulphuric acid, and an anode, which is a small platinum wire projecting into the electrolyte from a porcelain tube. Often called a Wehnelt interrupter after the inventor.

electrolytic lightning a r r e s t o r : Lightning arrestor consisting of an electrolyte, which covers two electrodes immersed in it with a film. This breaks down under a lightning discharge.

electrolytic rectifier: D e v i c e for changing an alternating current to a direct current by passing it through an electrolyte in which electrodes are immersed. The device acts as a "valve" to allow current to pass in one direction only.

electromagnet: Soft iron core having a coil wound around it through which an electric current is passed. The core is magnetized while the current flows, but is demagnetized when the current stops.

electromagnetic attraction: Attraction between opposite poles of an electromagnet.

electromagnetic brake: Brake used on car wheels and operated by electromagnets.

electromagnetic field: Space around a conductor or instrument, traversed by the electromagnetic waves set up by current in the conductor.

electromagnetic induction: Electric current set up in a conductor cutting the field of flux of an electromagnet.

electromagnetic repulsion: Repulsion between like poles of an electromagnet.

electromagnetic unit: Unit or standard of measurement of electromagnetic effects.

electromagnetic vibrator: Mechanical interrupter operated by an electromagnet.

electromagnetic wave: Form of electromagnetic energy radiated from a conductor and theoretically assuming the form of a wave. The rate of travel of these waves is approximately 186,000 miles per second.

electromagnetism: Science dealing with electricity and magnetism and their interaction.

electrometallurgy: Branch of metallurgy dealing with the use of electric currents either for electrolytic separation and deposition of metals from solutions, or with the utilization of electricity for smelting, refining, welding, annealing, etc.

electrometer: Device for measuring small voltages.

electromotive force: Electrical pressure or voltage which forces an electric current through a circuit.

electron: Electrical particle, of negative polarity.

electron theory: Theory that all matter consists of atoms which in turn comprise a positive nucleus and a number of negative electrons, which may be detached from the atom under certain conditions, leaving it positively charged.

electro-negative: Having a negative polarity.

electropathy: Science dealing with the use of electricity for medical purposes.

electrophorous: Device consisting of a disc of ebonite or similar substance, a metal plate and an insulator, used to produce an electric charge by induction.

electropism: Science dealing with the stimulation of vegetable growth by means of electricity.

electroplating: Process of covering a metal article with a metal deposit taken from an electrode and conveyed by an electrolyte in which the article is submerged.

electro-positive: Having a positive electrical polarity.

electro-receptive device: Device that receives electricity for its operation.

electroscope: Device that indicates the presence of a very small charge of electricity. It consists of a glass bottle having an electrode, which holds two strips of light foil. These attract or repel each other, depending upon the nature of the charge.

electrostatic: Pertaining to static electricity, or electricity at rest.

electrostatic capacity: Capacity to hold an electric charge, which is measured in farads and microfarads.

electrostatic field: Range around conductors, electrical machines and instruments where electrostatic effects take place.

electrostatic galvanometer: Galvanometer operated by the effect of two electric charges on each other.

electrostatic machine: Device which produces high-potential charges of static electricity by means of friction.

electrotherapeutics: Science dealing with the use of electric currents for curing diseases.

electrothermal: Pertaining to the heating effect of electric currents, and to electric currents produced by heat, as in thermo-couple.

electrotype: Metal plate used for printing. It is made by depositing metal on a form by means of electroplating.

element: (a) One of the parts to which all matter can be reduced. (b) One of the parts constituting a device, as a radio-tube element. (c) The resistor of an electrical heating device.

elevator cable: Flexible cable conveying electricity to an elevator. Also one of the cables supporting an elevator.

"Elexit": Trade name for certain standardized interchangeable fixture receptacles and plugs.

emissivity: Rate at which particles of electricity or heat are radiated from an object.

empire cloth: Cotton or linen cloth coated with linseed oil, and used as an insulator.

enameled wire: Wire having a coating or enamel baked on, which serves as insulation.

enclosed fuse: Fuse inside of a glass tube to prevent ignition of gas or dust.

end cell: One of a number of cells at the end of a storage battery, which can be cut in or out of the circuit to regulate the voltage.

end play: Distance of movement of a shaft in line with its length.

end thrust: Thrust exerted in line with the length of a shaft.

Endosmosis: The flow of a thin liquid to a denser liquid through a permeable partition.

energize: To put energy into; e.g., magnetizing an iron core of an electromagnet by passing a current through the coil.

energy: Capacity for performing work.

entrance switch: Switch to which the wires entering a building are connected.

equalizer: Connection between generators in parallel to equalize their voltage and current.

equalizing charge: Slight overcharge on a storage battery to raise the reading of the cells having the lowest specific gravity.

equator of magnet: Position halfway between the opposite poles of a magnet.

equipotential: Having the same potential.

equilibrium: State of rest or balance between two opposite forces, produced by their counter-action.

ether: Hypothetical element filling space to permit the passage of heat, light, electricity, gravity, etc., between solar bodies.

Evaporator: Heating device for evaporating water.

excite: To send a current through the field windings of a generator to set up a magnetic flux.

exciter: Small battery of generator furnishing current for the field windings of a large generator.

exciting current: Current which passes through the field windings of a generator.

extension: Length of cable or lamp-cord fitted with a plug and a socket to extend a lamp or other electric device further than the original point.

exploring coil: Device used for the detection of faults in underground cables. It consists of a coil and telephone receiver or head set, a current being induced in the coil at the point of leakage and causing a noise in the receiver. (b) Coil used to locate underground metals.

"Extra Best Best" iron wire: Trade name for the highest grade iron telegraph and telephone wire. Abbreviation EBB.

external circuit: A circuit entirely outside of the source of supply.

F

F: Abbreviation for frequency.

Fabrikoid: Trade name for a substitute for leather.

factor: Any one of the elements that contribute to produce a result.

factor of safety: Multiplier used in machine and structure design, designating the overload or safety capacity. E.g., a pressure vessel designed to withstand a pressure of 10 pounds per square inch may actually withstand 50 pounds per square inch, and the factor of safety is 5.

fading: Temporary diminution of signal strength in radio reception, due to atmospheric conditions.

Fahrenheit: A thermometer scale so graduated that the freezing point of water is 32° and its boiling point is 212°.

fall of potential: Drop in voltage between two points of an electric circuit.

false resistance: Resistance of counter e.m.f.

fan motor: Motor operating a fan.

farad: Unit for measuring electrical capacity. It is the capacity of a condenser which will give a pressure of one volt when a one-ampere current flows into it for one second.

farm-lighting generator: Small, gasoline-driven generator, producing current for farm light and power; usually a 32-volt and 2 or 3 kilowatt unit.

fathom: Nautical measure of length, equal to 6 feet. This unit is used to measure cables.

fault: Trouble in an electrical circuit.

fault finder: A resistance bridge for locating faults in telephone and telegraph circuits.

fault resistance: Resistance caused by a fault.

Fauré plate: Storage-battery plate consisting of a lead grid filled with paste.

feeder: Line supplying all the branch circuits with the main supply of current.

feeder box: Box into which the feeder is run for connection to a branch circuit.

fender: Device attached to street cars and other vehicles to pick up or brush aside obstacles.

ferro-manganese: Containing iron and manganese.

ferro-nickel: Containing iron and nickel.

fibre: A hard, tough insulating substance.

fibre cleats: Cleats made of fibre, used for holding conductors on flat surfaces.

fibre conduit: Insulating tubing made of moulded fibre.

field: Space occupied by the flux of a magnet.

field coil: Coil or winding around the field magnets of a generator or motor.

field discharge resistance: A resistance coil connected across the field winding of a generator permitting the winding to be discharged without a dangerous rise in voltage when the field circuit is opened by a switch. It is usually connected to a special d.p.d.t. knife switch having an auxilliary blade which connects the resistance just before the current to the winding is cut off.

field distortion: Variation of magnetic flux from the straight path between opposite poles in a generator, which is caused by armature reaction.

field flux: Space occupied by the lines of force of a generator field.

field intensity: Density of the field flux of a generator.

field magnet: The iron parts of a generator frame through which the flux of the coils concentrates.

field rheostat: A variable resistance device connected in the field circuit to control the voltage of a generator and the speed of a motor.

field winding. Coil on a field pole of a generator.

filament: Small wire in a lamp, which becomes white hot when electric current is passed through it.

film cutout: Insulating film between the two opposite wires inside of a lamp. The film burns out when the filament breaks and this permits the two wires to make contact, which provides a path for the current so that it can flow to other lamps, connected in series. These will not light unless the current passes through the defective lamp.

filters: Devices having inductance and capacity, and designed to suppress certain electrical frequencies.

fire-alarm systems: Apparatus which gives alarm in case of fire. Some of these systems consist of electrical circuits which, when closed automatically or otherwise, sound the alarm.

fire extinguisher: Devices using a liquid or powder to extinguish fire. They are used in power houses where there is danger of burning insulation on cables. Fire extinguishers used for this purpose must contain non-conducting liquids such as carbon tetrachlorid.

fish paper: Strong paper used for insulation.

fish wire: Flat, narrow, flexible, steel wire which is used to pull conductors through lengths of conduit.

fixed resistance: Non-Adjustable resistance.

fixture: Device for holding electric lamps, which is wired inside and is securely attached to the wall or ceiling.

fixture wire: Insulated, stranded wire used for wiring fixtures.

flaming arc: Arc which gives different colors due to impregnating the carbons with various salts and minerals.

flaming of arc: A flame bridging the gap between two carbons, instead of a steady arc, caused by the carbons being too far apart.

flasher: Automatic or motor-driven switch or series of switches for lighting electric signs intermittently.

flashing over: Passage of sparks from commutator segments traveling away from the brush, to the edge of the brush, which is then touching a segment adjacent to the one from which the spark originates.

flashlight: (a) Small, portable, electric light operated on one or more dry cells. (b) Trade name for an electric alarm clock.

flat-compound generator: Compound-wound generator having windings which give a constant voltage under different loads and speeds.

Fleming's rules: Rules for finding the direction of a conductor's motion through a magnetic field, the direction of the lines of force, and the direction of current flow through a conductor, applicable to direct current. Rule for Generators: Hold the thumb, the index finger, and the middle finger of the right hand so that they are at right angles to each other. The thumb will then point in the direction of the motion of the conductor, the index finger will point in the direction of the lines of force, and the middle finger will point in the direction of the current through the conductor. Rule for Motors: Hold the thumb, the index finger, and the middle finger of the left hand at right angles to each other. The thumb will then point in the direction of the motion of the conductor, the index finger will point in the direction of the lines of force, and the middle finger will point in the di-

rection of current through the conductor.

flexible cable: Cable consisting of insulated, stranded or woven conductors.

flexible conduit: Non-rigid conduit made of fabric or metal strip wound spirally.

flexible cord: Insulated conductor consisting of stranded wire.

floating battery: Storage battery connected in parallel with a generator and the load, so that the battery will consume the surplus current from the generator if the load is small, and will supply additional current if the load exceeds the output of the generator.

flood lights: Battery of lamps of high brilliancy, equipped with reflectors to supply a strong light.

flow: Passage of a current through a conductor.

fluctuating current: Current which changes in voltage and amperage at irregular intervals.

flush receptacle. Type of lamp socket, the top of which is flush with the wall into which the socket is recessed.

flush switch: Push-button or key switch, the top of which is flush with the wall into which it is recessed.

flux: Magnetic lines of force existing between two opposite magnetic poles.

flux density: Number of lines of force in a given cross-sectional area, which is measured in gausses.

focus: The point where rays of light, heat, sound, etc., meet after being reflected or refracted.

foot candle: See "candle foot."

foot-pound: Unit for measuring work. It is the energy required to raise a weight of one pound through a distance of one foot.

force: Energy exerted between two or more bodies which tends to change their relative shape or position.

form factor: Ratio of effective value of one half of a cycle of an alternating current to the average value of similar half cycles.

formers: Forms used for producing a number of windings of the same shape.

form-wound coils: Coils or windings built up on formers before they are placed in their proper position on armatures, field poles, etc.

forming battery plates: Passing an electric current through a storage battery to deposit peroxide of lead and spongy lead on the plates which makes them active.

Foucault currents: Same as Eddy currents.

four-pole: (a) Having four poles, as in a generator. (b) Having four contacts, as in a four-pole switch.

four-way switch: Switch that controls the current in four conductors by making or breaking four separate contacts.

four-wire, three-phase system: Distribution system having a 3-phase star connection, one lead being taken from the end of each winding and the fourth from the point where they are all connected together.

fractional pitch: Term used when the number of slots between the sides of an armature coil is not equal to the number of slots of each pole.

franchise: Permit from municipal, state, or national government to use public property, such as streets, for special purposes, as the installation of street-car lines.

frequency: Number of cycles or vibrations per second.

frequency changer: Motor-generator driven by an alternating current of one frequency and delivering current of another frequency.

frequency convertor: Same as "frequency changer."

frequency indicator: Device showing when two alternating currents are in phase or have the same frequency.

frequency meter: Device showing the frequency of an alternating current.

friction tape: Tape coated with black adhesive compound, used as insulation on wire joints, etc.

frog: Fixture for street-car tracks or trolleys where one track or trolley branches off, permitting the car to be run from one track onto another.

full pitch: Term used when the number of slots between the sides of an armature coil is equal to the number of slots of each pole.

fuller cell: Primary electric cell having two electrolytes: sulphuric acid and water, and a bichromate solution. These are separated by a porous cup. A cone-shaped zinc electrode is immersed in the cup, which also contains the sulphuric-acid solution and an ounce of mercury, and a carbon electrode is immersed in the bichromate solution.

fundamental units: Basic standards of measurement.

fuse: Safety device to prevent overloading a current. It consists of a short length of conducting metal which melts at a certain heat and thereby breaks the circuit.

fuse block: Insulated block designed to hold fuses.

fuse clip: Spring holder for a cartridge-type fuse.

fuse cutout: Fuse which, when melted, cuts out the circuit.

fuse link: An open fuse, or a length of fuse wire for refilling fuses.

fuse plug: Fuse mounted in a screw plug, which is screwed in the fuse block like a lamp in a socket.

fuse strip: A length of ribbon fuse as distinguished from wire fuse.

fuse wire: Wire made of an alloy which melts at a comparatively low temperature.

G

G: (a) Abbreviation for gram. (b) Symbol for mho, the unit of conductivity.

gage: Device for measuring.

galvanized: (a) Affected by galvanic action. (b) Metal coated with zinc.

galvanometer: Device for measuring small currents and voltages.

gang switch: Two or more switches installed in one box or one holder.

gas-filled: Filled with a gas as, for instance, an ordinary electric lamp.

gasoline-electric: Pertaining to a machine consisting of a gasoline engine, a generator driven by the engine, and one or more motors to produce electric power.

gauge: Same as "gage."

Gauss: Unit of flux density equal to one maxwell per square centimeter.

gauze brush: Generator or motor brush made of copper gauze.

Geissler tube: Gas-filled tubes, with or without fluorescent liquids, solids, or both, which emit light of various colors when a high-frequency c u r r e n t is passed through them.

gelatine battery: Battery having a jelly-like electrolyte.

generator: Machine that produces electricity.

generator busbar: Conductors on power switchboards to which a generator is connected.

generator loss: Difference between power required to drive a generator and the power it delivers, which is always less than the power input.

generator output: Power delivered by a generator, measured in watts or kilowatts.

geographical equator: Imaginary line around the earth halfway between the poles.

german silver: Alloy containing copper, nickel, and zinc, which is used for making resistance wire.

gilbert: Unit for measuring magnetic force. One gilbert is the magnetic force which sends one maxwell of flux through a magnetic circuit having a reluctance of one oersted.

glaze: Smooth finish applied to porcelain insulators to close the pores in order to prevent the absorption of moisture.

gramme armature: A ring type of armature.

graphite: A form of soft carbon.

gravity cell: Primary electric cell having two electrolytes, copper sulphate, and sulphuric-acid solutions, which are separated by gravity. The electrodes are zinc and copper.

gravity drop: A shutter or plate of an annunciator which, when released from a catch, drops by gravity.

Greenfield conductor: Flexible cable having a spirally wound metal covering.

Grenet cell: Primary electric cell of which the electrolyte is a solution of bichromate of potash in a mixture of sulphuric acid and water, and the electrodes are zinc and carbon. The z i n c electrode is lifted out of the electrolyte when the cell is not in use.

grid: (a) Frame of a storage-battery plate having spaces in which the paste is pressed. (b) An element of a vacuum tube which controls the rate of electron emission from the filament to the grid.

grid condenser: Small fixed condenser inserted in the line connecting with the grid of a vacuum tube used as a detector.

grid leak: Resistance shunted across a grid condenser in a radio circuit to allow dissipation of an excessive charge on the grid.

ground: See "earth."

ground circuit: Part of an electric circuit in which the ground serves as a path for the current.

ground clamp: Clamp on a pipe or other metal conductor connected to the ground for attaching a conductor of an electrical circuit.

ground detector: Device used in a power station to indicate whether part of the circuit is accidentally grounded.

ground indicator: Same as "ground detector."

ground plate: Metal plate buried in moist earth to make a good ground contact for an electrical circuit.

ground return: Ground used as one conductor of an electrical circuit.

ground wire: Conductor connecting an electrical device or circuit to the ground.

grounded neutral wire: The neutral wire of a 3-way distribution system which is connected to the ground.

grounded primary: Primary circuit of an induction c o i l or transformer connected to the ground.

grounding brush: Brush for making a ground connectior tr ꞓ moving part.

Grove cell: Primary electric cell which is similar to a Bunsen cell but has a platinum electrode instead of a carbon electrode.

growler: Coil around an iron core which is placed in contact with the core of an armature. When an alternating current or a pulsating direct current is passed through the growler coil, it magnetizes the core, which in turn induces a current in the armature winding. The purpose is to show whether a short circuit exists in the armature coil.

gutta percha: Hardened sap of a tropical tree which has high insulating quality and great resistance to destructive agencies such as water.

guy: A wire, rope, chain, or similar support for a structure such as a telephone pole, radio mast, etc.

H

H.: An abbreviation or symbol for intensity of magnetism.

h.p.: Horsepower.

hand advance: A device for controlling the advance and retard of the sparks in the ignition system.

hand regulation: Controlling the current or voltage by means of a hand operated device.

hard drawn copper: A method of producing high-grade copper of good mechanical strength.

hard fiber: A material made from a number of sheets of paper compressed tightly together. A good insulator.

hard rubber: An electrical insulation made by vulcanizing rubber.

harmonic currents: A series of currents which have frequencies that are multiple of the main current.

heat coil: A small coil placed on a telephone circuit to protect it from stray currents.

heat loss: The energy lost in a conductor due to its resistance.

heat run: A test made on a generator or motor to determine the amount of heating that takes place.

heating unit: That part of a heating appliance through which the current passes and produces heat.

head guy: A cable or wire fastened near the top of a pole to hold it in place.

head light: A light placed on the front end of a moving vehicle.

helix: A coil of wire; a solenoid.

henry: The electrical unit of inductance.

Hertzian wave: A radio wave.

high frequency: An alternating current that has many thousand cycles or alternations per second.

high potential: A high pressure or voltage, usually about six hundred volts.

high tension: A term used to refer to high voltage.

high tension magneto: A magneto used for ignition work in which the high-voltage current is produced in the magneto generator without the use of a separate induction coil.

holding magnet: An electromagnet used to hold metal objects while work is being done on them.

Holtz machine: A static electricity machine.

holophane: An electrical lighting globe with special surface for diffusing light.

homopolor generator: A generator having poles of one magnetic polarity only, instead of having alternate north and south pole.

hook-switch: A switch and hook on a telephone which is operated by placing or removing the receiver from that hook.

horizontal candle-power: The amount of light given off by a lamp measured in a horizontal direction from the light.

horn gap: A gap which is narrow at the bottom and widens out towards the top.

horsepower: The unit of power or work. An electrical horsepower is equal to 746 watts.

horsepower hour: The amount of power performed by 746 watts per one hour.

horseshoe magnet: A magnet bent in the shape of the letter U, or horseshoe.

hot conductor: A term used to refer to a conductor or wire which is carrying a current or voltage.

hot-wire meter: A meter which obtains a reading by the expansion of the length of wire or metal through which current flows.

howler: A device used in a telephone exchange to cause a noise in the receiver to indicate to the customer that the receiver has been left off the hook.

humming: A noise caused by the rapid magnetizing and demagnetizing of the iron core of a transformer, motor, or generator.

hunting: A condition in an electrical circuit where one machine tends to oscillate or run faster than another, and then run slower.

hydraulic: Pertaining to water or fluids in motion.

hydroelectric: The production of electricity by water-power.

hydrometer: An instrument or device which shows the specific gravity of a liquid as compared to water.

hysteresis: The tendency of magnetism to lag behind the current that produces it.

hysteresis curve: A curve that shows the relation between the magnetizing current and the amount of magnetism produced by it.

hysteresis loss: The heat produced by repeatedly magnetizing and demagnetizing the iron core of a machine.

I

I: An abbreviation for amperes of current.

I.E.S.: Illuminating Engineering Society.

i.h.p.: Indicated horsepower.

I-beam: A steel beam made in the form of a capital I.

Idle coil: A coil which does not produce any voltage or through which no current flows.

ignition: The igniting of a combustible charge in the cylinder of a gas engine.

ignition battery: A battery used to furnish the ignition current for an automobile engine.

ignition coil: An induction coil that produces a high-voltage current which jumps the gap in a spark-plug and ignites the charge in an automobile engine.

ignition distributor: A device that connects the proper spark-plug to the high-tension current at the right time in an automobile engine.

ignition generator: A generator used to produce the ignition current for an automobile engine.

ignition spark: The spark that passes between the gaps of the spark-plug, inside an automobile cylinder.

ignition switch: A switch that is used for turning on and off the primary ignition coil.

ignition timer: A device that closes and opens the primary circuit of an induction coil at the proper instant, to produce a spark in a cylinder of an automobile engine.

illumination: The directing of light from its source to where it can be used to the best advantage.

impedance: The apparent resistance of a circuit to alternating current. It is composed of resistance and reactance.

impedance coil: A reactance or choke coil, used to limit the flow of current.

impregnated cloth: A cotton cloth that has been saturated with insulating varnish and dried.

impressed voltage: The voltage or pressure acting upon any device.

impulse: A sudden change, such as an increase or decrease in voltage or current.

incandescent lamp: A lamp in which light is produced by the heating of a small filament inside of a glass bulb.

inclined coil instrument: A voltmeter or ammeter in which the coil or moving vane are inclined in relation to the pointer.

incomplete circuit: An open circuit.

india rubber: A soft rubber used to insulate or cover electrical conductors and wires.

indicated horsepower: The horsepower determined by calculation taken from an indicator diagram.

indicating switch: A switch that shows whether it is turned "ON" or "OFF."

indirect lighting: Light that is thrown against a ceiling having a light colored surface and reflected and diffused in the room being lighted.

induced current: Current that is produced by inductance from another circuit.

induced e.m.f.: A voltage that is produced by induction from another circuit.

induced magnetism: Magnetism that is produced by electric current or by the action of other magnetism.

induced voltage: A voltage or pressure produced by induction.

inductance: The ability of an electric circuit to produce induction within itself.

inductance coil: A coil connected in an electric circuit in order to increase the resistance of that circuit to alternating current.

induction: The influence exerted by a magnet or magnetic field upon conductors.

induction coil: A coil used to produce a high-voltage. It consists of two windings placed on an iron core. The voltage is produced by stopping quickly the flow of current in the coil.

induction furnace: An electric furnace in which the metal forms a secondary circuit of the transformer and is heated by current flowing through the metal.

induction generator: An induction motor, operated about synchronous speed, which produces an electric current.

induction meter: A meter used on alternating current in which rotation of a disk is caused by the magnetic lines of force produced by a current and a voltage coil passing through the disk.

induction motor: An alternating-current motor which is operated by induced magnetism from the winding placed on the stator. It does not operate at synchronous speed.

induction regulator: A transformer in which the voltage produced in a secondary winding is varied by the changing of position of the primary winding.

inductive load: The load connected to an alternating-current system which causes the current to lag behind the voltage.

inductive reactance: The reactance produced by self-inductance.

inductive resistance: The apparent resistance that is caused by self-induction in a circuit.

inductor: That part of an armature winding which lies entirely on one side of the armature coil, and in which a voltage is produced.

industrial controller: A device or rheostat for controlling the speed of electric motors.

inertia: The tendency of a body to remain at rest or in motion at the same speed.

initial voltage: The pressure at the start; as the voltage at the terminal of a storage battery when it is placed on change; that voltage which causes the appearance of corona around an electric conductor.

input: All of the power delivered to an electric device or motor.

inside wiring: The wiring inside of a residence or building.

installation: All of the electrical equipment or apparatus used in a building including the wiring.

instrument transformer: A transformer used to change the voltage or current supplied to meters.

insulate: To place insulation around conductors or conducting parts of a device or object.

insulating: The placing of insulation around electrical conductors.

insulating compound: An insulating wax which is melted and poured around electrical conductors in order to insulate them from other objects.

insulating joint: A thread or coupling in which the two parts are insulated from each other.

insulation resistance: The resistance offered by an insulating material to the flow of electric current through it.

insulating varnish: A special prepared varnish which has good insulating property and is used to cover the coils and windings on electric machines and improve the insulation.

insulator: A device used to insulate electric conductors.

intake: A place where air or water enters a machine, tunnel, or pipe.

integrating meter: A meter that keeps the record of the total amount of power, current, etc., that passes through it in a given time.

intensity: The intensity of the current is the number of amperes that flows through a conductor in a given time.

intercommunicating telephone: A telephone system that connects up to the several offices in the same building or plant without the use of a central operator.

interior wiring: Wiring placed on the inside of buildings.

intermittent current: A current, that starts and stops its flow at regular intervals.

intermittent rating: When a machine is operated for a short time only and allows a long period of rest, it has an intermittent rating.

internal circuit: The circuit formed inside a device or machine.

internal resistance: The resistance of the winding of an electrical machine, or between terminals of a primary cell or a storage battery.

internal short-circuit: A short-circuit occurring between the positive and negative plates in a storage battery due to a defective separator.

internal wiring: The wiring inside of a device or a machine.

interpoles: Magnetic poles placed between the main poles of a motor or generator.

interrupter: A device that opens or closes a circuit many times per second.

interrupter contact: The contact where a circuit is broken by an interrupter.

interrupter gap: The greatest amount of distance or space between the contacts of an interrupter.

invar: A resistance wire composed of nickel and steel.

inverse ratio: A ratio where one value increases and the other value decreases.

inverted converter: A rotary or synchronous converter which changes direct current into alternating current.

ion: The two minute parts into which a molecule is divided when it is separated into its elements.

I²R loss: The power loss due to the current flowing through the conductor which has resistance. This loss is converted into heat.

iron loss: The hysteresis and eddy current losses in iron cores of electric machinery.

ironclad armature: An armature in which the windings are placed in slots cut in the armature core.

ironclad magnet: A magnet which has an iron core extending around the outside of the coil and through which the magnetism flows.

isolated plant: An electric light plant used to furnish power for a small community or a few firms, and the power of the plant is not sold to the public.

J

J: Abbreviation for joule.

jack: The terminal of two telephone lines on a switchboard of a telephone exchange.

joint: The uniting of two conductors by means of solder.

joint resistance: The combined or total resistance of two or more resistances connected in series or parallel.

joule: A unit of electrical work. A current of one ampere flowing through a resistance of one ohm for one second.

journal: That part of a shaft that turns or revolves in the bearings.

jump spark: A spark that passes between two terminals or across a gap. It is produced by high voltage.

jumper: A temporary connection made around part of a circuit.

junction box: A box in a street distribution system where one main is connected to another main; also a box where a circuit is connected to a main.

K

K.: Abbreviation or symbol for dielectric constant.

k.w.: Abbreviation for kilowatt.

kaolin: A kind of clay used in making porcelain insulators.

keeper of magnet: A bar of soft-iron placed across the poles of a magnet when it is not being used.

Key: A device for opening and closing a circuit by moving a lever. It is used in telephone and telegraph apparatus.

key switch: A switch for turning on and off electric circuits which are operated by means of a special key.

key socket: A socket with a device that opens and closes the circuit, thus turning the lamp off or on.

keyless socket: A socket which does not have a key or device for turning on or off the lamp.

kicking coil: A reactance or choke coil.

kilo: A prefix when placed before a word means 1000 times that indicated by the word.

kiloampere: One thousand amperes.

kilovolt: One thousand volts.

kilowatt: One thousand watts.

kilowatt-hour: One thousand watt-hours.

knife switch: A switch that has a thin blade that makes contact between two flat surfaces or short blades to complete the circuit.

knob insulator: A porcelain knob to which electric wires may be fastened.

L

L.: An abbreviation for length.

lag: To drop behind.

lag of brushes: The distance the brushes are shifted on a motor or generator in order to prevent sparking.

lagging coil: A small coil used in alternating watt-hour meter to compensate for the lagging current in the voltage coil.

lagging current: The lagging of the current behind the voltage wave in an inductive alternating-current system.

laminated: Built up out of thin sheets or plates which are fastened together.

laminated core: A core built up of thin soft iron sheets placed side by side and fastened together.

laminations: One of the plates used in building a laminated core.

lamp: A device used to produce light.

lamp bank: A number of incandescent lamps connected in series or in parallel and used as resistances.

lamp base: The metal part of an incandescent lamp which makes contact with the socket.

lamp bulb: A term used in referring to an incandescent electric lamp.

lamp circuit: A branch circuit supplying current to lamps only, and not to motors.

lamp cord: Two flexible stranded insulated wires twisted together and used to carry the current from the outlet box to the lamp socket.

lamp dimmer: An adjustable resistance connected in a lamp circuit in order to reduce the voltage and the brightness of the lamps.

lamp socket: A receptacle into which the base of the lamp is inserted, and which makes connection from the lamp to the circuit.

lap winding: An armature winding in which the leads from the coil to the commutator lap over each other.

lap-wound armature: An armature that has a lap winding.

lateral: A conduit that branches off to the side from the main conduit.

lava: A kind of stone that has insulating properties.

lead (pronounced lĕd): An acid resisting metal that is used in making parts for storage batteries.

lead battery: A storage battery in which the plates are made from lead.

lead burning: The process of uniting two pieces of lead together by melting the edges.

leads (pronounced leeds): Short lengths of insulated wires that conduct current to and from a device.

lead of brushes: The distance that the brushes are moved on the commutator of a generator or motor to prevent sparking.

leading current: When the current of an alternating-current system reaches its maximum value before the voltage does, it is called a leading current.

leading-in wires: Wires used to carry current from the outside of buildings to the inside of buildings.

leak: A loss of charge in a storage battery where current can flow through a circuit, or to ground, due to defective insulation.

leakage flux: Lines of force or magnetism that do not flow through the path intended for them but take another path and do not do any useful work.

Leclanche cell: A primary cell which uses carbon and zinc rods or plates for electrodes.

left-hand rotation: A shaft or motor that revolves in a counter clockwise direction; that is, opposite to that of the hands of a clock.

Leyden jar: A glass jar covered inside and out with a thin metal covering, and used as a condenser.

lifting magnet: An electromagnet used to lift iron and steel objects.

light load: A load that is less than the usual or normal load on the circuit.

lighting fixture: An ornamental device that is fastened to the outlet box in the ceiling and which has sockets for holding the lamps.

lighting transformer: A transformer that is used to supply a distribution circuit that does not have motors connected to it.

lightning arrester: A device that allows the lightning to pass to the ground thus protecting electrical machines.

lightning rod: A rod that is run from the ground up above the highest point of a building.

limit switch: A switch that opens the circuit when a device has reached the end of its travel.

line of force: An imaginary line which represents the direction of magnetism around a conductor or from the end of a magnet.

line drop: The loss in voltage in the conductors of a circuit due to their resistance.

line insulator: An insulator for use on an overhead transmission line.

line reactance: The reactance in the transmission line or conductor outside of the supply station.

line resistance: The resistance of the conductor forming the transmission line.

lineman: A man who erects or works on an electric transmission line.

link fuse: A fuse that is not protected by an outside covering.

litharge: A compound made from lead used in the active material of storage battery plates.

live: A circuit carrying a current or having a voltage on it.

load: The work required to be done by a machine. The current flowing through a circuit.

load control: Changing the output of a generator as the changes of load occur on a circuit.

load dispatcher: A person who supervises or controls the amount of load carried by the generating station on a system.

load factor: The average power consumed divided by the maximum power in a given time.

loading coils: Small coils placed in series with telephone lines in order to improve the transmission of speech.

local action: A discharge between different parts of a plate in a storage battery or primary cell caused by impurities in the parts used.

local current: An Eddy current.

locked torque: The twisting or turning power exerted by a motor when the rotating part is held stationary and normal current supplied to the winding.

lodestone: Magnetic iron ore.

log: A record of events taken down as they occur.

long shunt: Connecting the shunt across the series field and armature, instead of across the armature terminals.

loop circuit: A parallel or multiple circuit.

loop test: A test using the Wheatstone bridge, and a good line to locate an accidental ground on a line.

loose contact: A poor connection that does not make proper contact.

loud speaker: An electrical device that reproduces sound loud enough to be heard across a room.

low frequency: A current having a small number of cycles per second.

low potential: A system where the voltage between wires is usually less than 600 volts.

low tension: Low pressure or voltage.

low tension winding: The winding on a transformer which produces or has the lowest voltage.

low voltage release: A device that opens the circuit when the voltage drops down to a certain value, for which it is adjusted.

lugs: Terminals placed on the end of conductors to enable the wire to be attached or detached quickly.

lumen: The unit of electric lighting.

luminarre: An ornamental electric lighting fixture.

luminosity: In electric lighting work it is the brightness of a color compared with light.

luminous flux: In lighting work it is the amount of light directed down toward the point where it can be used.

M

M: A symbol of mutual induction the unit for which is a henry.

M.C.B.: Master Car Builder.

M.D.F.: Main distributing frame in a telephone exchange.

m.f.d.: Microfarad.

M-G: An abbreviation for motor-generator sets.

m.p.h.: Miles per hour.

machine rating: The amount of load or power a machine can deliver without overheating.

machine switching: A telephone exchange where the connections from one party to another are made by a machine instead of by an operator.

magnet: A body that will attract iron or steel.

magnet charger: A large electro-magnet used to magnetize permanent magnets.

magnetic coil: The winding of an electromagnet.

magnet core: The iron in the center of the electromagnet.

magnet winding: The wire wound on a spool, forming an electromagnet.

magnet wire: A small single conductor copper wire insulated with enamel, cotton, or silk, used in winding armatures, field coils, induction coils, and electromagnets.

magnetic attraction: The pull or force exerted between two magnets or between magnets and an iron or steel body.

magnetic blow-out: A magnet arranged so that the arc between contacts is quickly lengthened and extinguished.

magnetic brake: A friction brake which is applied or operated by an electromagnet.

magnetic bridge: An instrument that measures the permeability and reluctance of magnetic material.

magnetic circuit: The paths taken by lines of force in going from one end of the magnet to the other.

magnetic compass: A small magnetized needle which indicates north and south directions.

magnetic contactor: A device, operated by an electromagnet, which opens and closes a circuit.

magnetic density: The amount of magnetism or magnetic lines of force per square inch or centimeter.

magnetic dip: The angles that a balanced needle makes with the earth when it is magnetized.

magnetic equator: An imaginary line joining the points about the earth where the compass needle does not have any dip.

magnetic field: The magnetic lines of force that pass in the space around a magnet.

magnetic flux: Magnetism or the number of lines of force in a magnetic circuit.

magnetic force: The attraction between magnetic poles or magnets, producing magnetism in a magnetic body by bringing it near a magnetic field.

magnetic lag: The tendency for magnetism to lag behind the current or force producing magnetism.

magnetic leakage: Lines of force that do not do useful work by passing through a path that is not in a working field.

magnetic lines of force: Magnetism about a conductor or flowing from magnet.

magnetic material: Materials which conduct lines of force easily—iron and steel.

magnetic needle: A small magnet that points in the direction of the magnetic lines of force about the earth.

magnetic pole: The ends of the magnet where the magnetism enters or leaves the magnet.

magnetic potential: Magnetic pressure which produces a flow of magnetic lines of force.

magnetic pulley: A pulley with an electromagnet inside of it, and used to separate iron and steel from other materials passed over it.

magnetic saturation: The greatest number of magnetic lines of force or magnetism that a body or substance can carry.

magnetic screen: A soft iron body around which magnetism is conducted instead of going through the center of that object.

magnetic shunt: A definite path for magnetic lines of force to pass through instead of the main path.

magnetic switch: A switch that is operated or controlled by an electromagnet.

magnetism: That invisible force that causes a magnet to attract iron and steel bodies.

magnetite: Magnetic iron ore.

magnetization curve: A curve that shows the amount of magnetism, expressed in lines of force, produced by a certain magnetizing force.

magnetize: To cause a substance to become a magnet.

magnetizing force: That force which produces magnetism. It is measured in ampere-turns.

magneto: A small generator that has a permanent field magnet.

magneto ignition: Igniting the charge in a combustion engine from a magneto generator.

magnetomotive force: That force which produces magnetism; it is expressed in ampere-turns.

main: The circuit from which all other smaller circuits are taken.

main feeder: A feeder supplying power from the generating station to the main.

make-and-break ignition: Igniting the charge in an internal combustion engine by the spark produced when contacts carrying current are opened.

maintenance: Repairing and keeping in working order.

manhole in conduit: An opening or chamber placed in a conduit run large enough to admit a man to splice or join cables together.

manganese steel: An alloy of steel having a large percent of the metal called manganese.

manual: Operated by hand.

mariner's compass: A compass used by sailors for directing the course of a ship.

master switch: A switch that controls the operation of other switches or contact switches.

maximum demand: The greatest load on a system occurring during a certain interval of time.

maximum demand meter: A meter that registers or indicates the greatest amount of current or power passing through a circuit within a given time.

maxwell: A unit of magnetic flux or lines of force.

mazda lamp: A certain trade name for an incandescent lamp using a tungsten filament.

mean horizontal candle-power: The average candle-power measured on a horizontal plane in all directions from the lamp filament.

mean spherical candle-power: The average candle-power of a lamp measured in all directions from the center of the lamp.

meg or mega: A prefix that means one million times.

megger: An instrument that measures the resistance in megohms.

megohms : A resistance of one million ohms.

mercury: A silvery white metal liquid; often called quicksilver.

mercury-arc rectifier: A rectifier in which alternating current is changed to direct current by the action of mercury vapor on electrodes.

mercury vapor lamps: The lamps or lights in which light is produced by passing a current through mercury vapor.

mesh connection: A closed circuit connection in armature winding.

messenger wire: A wire used to support a trolley, feeders, or cable.

metal conduit: Iron or steel pipe in which electric wires and cables are installed.

metal moulding: A metal tube or pipe, installed on the ceiling or walls of a building, in which electric wires are installed.

metallic circuit: A circuit that uses wires to return the current to the starting point instead of returning it through the ground.

metallic filament: An incandescent lamp filament made from a metal such as tantalum or tungsten.

meter: A device that records and indicates a certain value of electricity.

meter loops: Short pieces of insulated wire used to connect a watthour meter to the circuit.

metric system: A system of weights and measures based upon a meter (39.37 inches) for length and a gram ($\frac{1}{28}$ ounce) for weight.

Mho: The reciprocal of the resistance of a circuit which is called conductivity.

mica: A transparent mineral substance used for insulating commutators.

mica undercutter: A tool used to cut the mica below the surface of the commutator segment.

micanite: A trade name for small pieces of flake mica cemented together with an insulating compound.

micro: A prefix meaning one-millionth part.

micro-ampere: T h e one-millionth part of an ampere. $\frac{1}{1,000,000}$ or .000001 amperes.

microfarad: One-millionth of a farad.

microhm: One-millionth of an ohm.

microphone: A telephone transmitter in which the resistance is varied by a slight change in pressure on it.

microvolt: One-millionth of a volt.

mil: One-thousandth part of an inch; $\frac{1}{1000}$ or .001 inch.

mile-ohm: A conductor that is one mile long and has a resistance of one ohm.

mil-foot: A wire that is one-thousandth of an inch in diameter and one foot long.

milli: Prefix to a unit of measurement, denoting one-thousandth part of it.

milli-ammeter: An instrument that reads the current in thousandths of an ampere.

milli-ampere: $\frac{1}{1000}$ or .001 amperes. One-thousandth of an ampere.

milli-henry: One-thousandth of a henry.

milli-volt: One thousandth of a volt.

milli-voltmeter: A voltmeter that reads the pressure in one-thousandth of a volt.

mineralac: A trade name of an insulating compound or wax.

miniature lamp: The smallest size of incandescent lamp that uses a screw threaded base.

mirror galvanometer: A very sensitive galvanometer with a mirror attached to the moving element which reflects a spot of light over a scale.

moment: That which produces motion.

monel metal: An alloy of nickel and copper that is not eaten away by acids.

momentum: The tendency of a moving body to remain in motion at the same speed.

molecule: The smallest existing particle of a compound substance.

moonlight schedule: A list showing the time to turn the street lights out one hour after the moon rises, and turn them on one hour before the moon sets.

Morse code: A series of dots and dashes as signals transmitted by telegraph used to transmit messages.

motor: A machine t h a t changes electrical energy into mechanical power.

motor converter: A form of rotary or cascade converter.

motor circuit: A circuit supplying current to an electric motor.

motor-generator: An electric motor driving a generator changing alternating to direct current or the reverse.

moulded insulation: A form of insulating material that can be placed in a mold and pressed into shape.

moulding: A wooden or metal strip provided with grooves to receive rubber covered electric wires.

moving coil meter: An electrical instrument of the d'Arsonval type which has a coil of fine wire moving between permanent magnets.

multiple: Connected in parallel with other circuits.

multiple circuit: A circuit in which the devices are connected in parallel with each other.

multiple series: A parallel connection of two or more series circuits.

multiple winding: A winding where there are several circuits in parallel.

multiple unit control: Controlling the operation of motors on several cars of an electric train from one point.

multiplex telegraphy: Sending one or more messages in both directions in the same circuit at the same time.

multiplex wave winding: A wave-wound armature that has more than two circuits in parallel.

multiplier: An accurately calibrated resistance connected in series with a voltmeter to enable it to be used on higher voltage circuits.

multipolar: Having more than two pole-pieces and field coils.

multi-speed: An electric motor that can be operated at several definite speeds.

mush coil: An armature coil that is not wound in regular layers.

N

N: A symbol used for revolutions per second or minute; often used to denote the North pole of a magnet.

N.E.C.: Abbreviation for National Electric Code; often called Underwriter's Code.

N.E.L.A.: Abbreviation for National Electric Light Association.

N.F.P.A.: Abbreviation for National Fire Prevention Association.

n.h.p.: Abbreviation for nominal horsepower.

name plate: A small plate placed on electrical machines which gives the rating of the machine and the manufacturer's name.

natural magnet: Magnetic ore or lodestone.

needle: A magnetized piece of steel which can be swung from the center and will point in the direction in which the magnetic lines of force are flowing.

needle point: The sharp point on a spark gap.

negative: The point towards which current flows in an external electrical circuit; opposite to positive.

negative brush: The brush of a generator out of which current enters the armature. In a motor the brush at which current leaves the armature.

negative charge: Having a charge of negative electricity.

negative conductor: The conductor that returns the current to the source after it has passed through a device and has been used.

negative electrode: The electrode by which the current leaves an electrolyte and returns to its source.

negative feeder: A feeder connected to the negative terminal on a generator to aid the current returning to the generator.

negative plate: The s p o n g e lead plate of a lead acid-battery. In a primary cell the terminals to which the current returns from the external circuit.

negative pole: The S-pole of a magnet. The pole that the lines of force enter the magnet.

negative side: That part of the circuit from w h e r e the current leaves the consuming device to where it re-enters the generator.

negative terminal: That terminal to which the current returns from the external circuit.

neon: An inert gas used in electric lamps.

nernst lamp: A lamp in which light is produced by passing the current or electricity through rare oxide contained in a tube.

network: A number of electrical circuits or distribution lines joined together.

neutral: Not positive or negative although it may act as positive to one circuit and negative to another.

neutral conductor: A middle conductor of a three-wire direct-current or single-phase circuit.

neutral induction: The variation of current in one circuit which causes a voltage to be produced in another circuit.

neutral position: That point on the commutator where the armature conductors do not produce any voltage, because they are not cutting lines of force at that point.

neutral terminal: A terminal which may be positive to one circuit and negative to another circuit.

neutral wire: That wire in a three-wire distribution circuit which is positive to one circuit and negative to the other.

nichrome: An alloy of nickel and chromium which forms a resistance wire that can be used at a high temperature.

nickel: A silver white metal.

nickel silver: An alloy of copper, zinc, and nickel.

nickel steel: An alloy steel containing a small per cent of nickel.

nitrogen lamp: An incandescent lamp containing nitrogen or other inert gas instead of a vacuum.

non-conductor: That material which does not easily conduct electric current; an insulator.

non-inductive: Having very little self-induction.

non-inductive load: A load connected to a circuit that does not have self-induction. With alternating-current circuit, the current is in phase with the voltage.

non-inductive winding: A winding arranged so that it does not have any self-induction.

non-magnetic: Materials that are not attracted by a magnet are called non-magnetic.

normal: The general or usual conditions for that particular device or machine.

North pole: The end of the magnet at which the lines of force leave it. The end of a freely suspended magnet that will point towards the North.

numerator: In fractions the word or number written above the horizontal line.

O

O.K.: An abbreviation which means all right.

oersted: The unit of magnetic reluctance which is the resistance of metal to the flow of magnetism through them.

ohm: The unit used to express the resistance of a conductor to the flow of electric current through it.

Ohm's law: A rule that gives the relation between current, voltage, and resistance of an electric circuit. The voltage (E) is equal to the current (I) in amperes times resistance (R) in ohms. The current (I) equals the voltage (E) divided by the resistance (R) of the circuit. The resistance (R) is equal to the voltage (E) divided by the current (I).

ohm-mile: A conductor a mile long and has a resistance of one ohm.

ohmic resistance: The resistance of a conductor due to its size, length, and material.

oil circuit breaker: A device that opens an alternating-current circuit in a tank of oil which extinguishes the arc.

oil switch: A switch whose contacts are opened in a tank of oil.

oiled paper: A paper treated with an insulating oil or varnish.

open circuit: A break in a circuit. Not having a complete path or circuit.

open circuit battery: A primary cell that can only be used for a short time, and requires a period of rest in order to overcome polarization.

open coil armature: An armature winding in which the ends of each coil are connected to separate commutator bars.

open delta connection. A transformer connection in which two single-phase transformers are used to form two sides of a delta connection.

open wiring: Electric wires fastened to surfaces by the use of porcelain knobs. Wiring that is not concealed.

ordinate: The vertical lines drawn at various points along the horizontal base line to indicate values on that base line.

oscillating discharge: A number of discharges obtained one after another from a condenser; each one is less than the one before.

oscillograph: A very sensitive and rapid galvanometer which shows changes occurring in electrical circuits.

outboard bearing: A bearing placed on the outside of a pulley of a machine.

outlet: A place where electrical wires are exposed so that one can be joined to the other.

outlet box: An iron box placed at the end of conduit where electric wires are joined to one another and to the fixtures.

output: The amount of current in amperes or watts produced by a generator or a battery.

overcompound: When the series field coils of a generator are designed so that the voltage will increase with an increase in load, the generator is said to be overcompounded.

overdischarge: Discharge from a storage battery after the voltage has dropped to the lowest normal discharge value.

overhead: Electric light wires carried out doors on poles.

overload: Carrying a greater load than the machine or device is designed to carry.

overload capacity: The amount of load beyond a rated load that a machine will carry for a short time without dangerously overheating.

overvoltage: A voltage higher than the normal or usual voltage.

ozone: A form of oxygen produced by electrical discharge through air.

P

P: Abbreviation for power.

P.B.X.: Private branch telephone exchange.

P.D.: Potential difference.

panel box: The box in which switches and fuses for branch circuits are located.

parabolic reflector: A reflector built in the form of a parabolic curve in order to reflect the light in a narrow beam.

paraffin: A wax used for insulating bell wire.

parallax: The difference caused by reading the scale and pointer of an instrument at an angle instead of straight in front of it.

parallel: Two lines extending in the same direction which are equally distant at all points. Connecting machines or devices so that the current flows through each one separately from one line wire to another line wire. Also called multiple.

parallel circuit: A multiple circuit. A connection where the current divides and part flows through each device connected to it.

parallel series: A multiple series. A number of devices connected in series with each other, forming a group; and the groups are connected in parallel with each other.

parallel winding: A lap armature winding.

paramagnetic: Material that can be attracted by a magnet.

para rubber: The best grade of india rubber.

pasted plate: A storage battery plate in which the active material is prepared as a paste and forced into openings in the grid.

peak load: The highest load on a system, or generator, occurring during a particular period of time.

peak voltage: The highest voltage occurring in a circuit during a certain time.

pendant switch: A small push button switch, hanging from the ceiling by a drop cord, used to control the flow of current to a ceiling light.

permanent magnet: A magnet that holds its magnetism for a long time.

permeability: The ease with which a substance conducts or carries magnetic lines of force.

permeability curve: A curve that shows the relation of the magnetizing force (ampere-turns) and number of lines of force produced through a certain material.

permeameter: An instrument used to test the permeability of iron and steel.

permittivity: The dielectric constant.

peroxide of lead: A lead compound used in making storage battery plates.

petticoat insulator: An insulator the bottom part of which is in the shape of a cone with the inside hollow for some distance.

phantom line: An artificial line over which messages can be sent the same as over an ordinary line.

phase: The fraction of a period of cycle that has passed since an alternating voltage or current has passed through zero value in the positive direction.

phase advancer: A machine used to improve the power factor of a system by overcoming the logging current.

phase angle: The difference in time between two alternating-current waves expressed in degrees. A complete cycle of 360 degrees.

phase converter: A machine that changes the number of phases in an alternating-current c i r c u i t without changing the frequency.

phase failure: The blowing of a fuse or an opening of one wire or line in a two- or three-phase circuit.

phase indicator: A device that shows whether two electric machines are "in step" or in synchronism.

phase rotation: The order in which the voltage waves of a three-phase circuit reach their maximum value, as ABC or ACB.

phase shifters: Devices by which power-factor can be varied on a circuit when testing meters.

phase splitter: A device that causes an alternating current to be divided into a number of currents that differ in phase from the original.

phase winding: One of the individual armature windings on a polyphase motor or generator.

phosphor bronze: Bronze to which phosphor has been added in order to increase its strength.

photometer: An instrument used to measure the intensity of light.

pig tail: Five braided copper wires used to connect the carbon brush to its holder.

pike pole: A small pole with a sharp spike in one end. It is used by wiremen in raising and setting wood poles.

pilot brush: A small brush used to measure the voltage between adjacent commutator bars.

pilot cell: A cell in a storage battery used as a standard in taking voltage and specific gravity readings.

pilot lamp: A small lamp used on switchboards to indicate when a circuit switch or device has operated.

pitch: The number of slots between the sides of an armature coil. The distance from a certain point on one to a like point on the next.

pith balls: Small balls made from the light soft spongy substance in the center part of some plants and corn cobs.

pivots of meters: The shaft to which the moving part of the meter is fastened and which turns on a bearing.

Planté plates: A storage battery plate in which the active material is formed by charging and discharging the battery many times.

plate condenser: A condenser formed by a number of plates with insulating material between them.

plating dynamo: A generator that produces a low voltage direct current for use in electroplating work.

platinum: A gray-white metal that is not easily oxidized and which makes good contact points.

platinum-iridium: An alloy of platinum and iridium, which is a harder metal than platinum.

plug: A screw thread device that screws into an electric light socket and completes the connection from the socket to the wires fastened to the plug.

pocket meter: A small voltmeter or ammeter mounted in a case that can be carried in the coat pocket.

polar relay: A relay that operates when the direction of the flow of current changes.

polarity: Being positive or negative in voltage, current flow, or magnetism.

polarity indicator: An instrument that indicates the positive or negative wires of a circuit.

polarity wiring: Using a white or marked wire for the ground side of a branch circuit.

polarization: The forming of gas bubbles on the plates of a primary cell which reduces the current produced by the cell.

polarized: Having a definite magnetic polarity.

polarized armature: The armature of a magnet that has a polarity of its own and which is attracted only when the direction of the flow of current in the windings produces a pole of opposite polarity.

pole: The positive and negative terminal of an electric circuit. The ends of a magnet.

pole changer: A device that changes direct current into alternating current.

pole piece: The end of the field magnet or electromagnet that forms a magnetic pole.

pole pitch: The number of armature slots divided by the number of poles.

pole shoe: A piece of metal having the same curve as the armature that is fastened to the field magnet of a generator or motor.

pole strength: The number of magnetic lines of force produced by a magnet.

pole tips: The edges of the field magnets toward and away from which the armature rotates.

polyphase: Having more than one phase.

polyphase circuit: A two- or three-phase circuit.

polyphase transformer: A transformer in which the windings of all the phases are located inside the same case or cover.

porcelain: A hard insulating material made from sand and clay which is molded into shape and baked.

porous cell: A porous jar used with primary cells that use two different electrolytes that must be kept separate.

portable instrument: A meter so designed that it can be moved from one place to another.

positive: The point in a circuit from which the current flows; opposite to negative.

positive brush: The brush of a generator from which the current leaves the commutator; the brush of the motor through which current passes to the commutator.

positive electricity: The kind of electricity produced by rubbing a glass rod with silk.

positive electrode: The electrode or terminal that carries the current into the electrolyte.

positive feeder: A wire or cable acting as a feeder that is connected to the positive terminal of a generator.

positive plate: The peroxide of lead plate in a lead-acid storage battery.

positive terminal: The terminal of a battery or generator from which the current flows to the external circuit.

potential: The pressure, voltage, or electromotive force that forces the current through a circuit.

potential coil: The voltage or pressure coil of a meter that is connected across the circuit and is affected by changes in voltage.

potential regulator: A device for controlling or regulating the voltage of a generator or circuit.

potential transformer: A transformer used to step the voltage down for voltmeters and other instruments.

potentiometer: An instrument used to compare a known or standard voltage with another voltage.

pothead: A flared out pot or bell attached to the end of a lead covered cable and filled with insulating compound.

poundal: The unit of force which, acting for one second, will give a body that has mass of one pound a velocity of one foot per second.

power: The rate of doing work. In direct current circuits it is equal to $E \times I$. The electrical unit is the watt.

power circuit: Wires that carry current to electric motors and other devices using electric current.

power factor: The ratio of the true power (watts) to the apparent power (volts $\times$ amperes). Cosine of the angle of lag between the alternating current and voltage waves.

power factor meter: A meter that indicates the power factor of the circuit to which it is connected.

power loss: The energy lost in a circuit due to the resistance of the conductors; often called I^2R loss.

power plant: The generators, machines, and buildings where electrical power or energy is produced.

practical units: The electrical units used in everyday practical work —the ohm, volt, ampere, watt, etc.

precision instrument or meter: A very accurate meter or instrument used in testing or comparing other meters.

press board: A hard smooth paper or cardboard used for insulation in generators and transformers.

pressure: The voltage which forces a current through a circuit; also called potential difference.

pressure wires: Wires going from the end of a feeder to a voltmeter in the power station.

primary: That which is attached to a source of power, as distinguished from the secondary.

primary cell: A cell producing electricity by chemical action, usually in acid acting on two different metallic plates.

primary circuit: The coil or circuit to which electric power is given and which transfers it to the secondary by induction.

primary winding: The winding which receives power from the outside circuit.

prime mover: An engine, turbine, or water wheel that drives or operates an electric generator.

prony brake: A friction brake or a pulley used as a dynamometer to measure the torque turning power of a shaft.

proportional: A change in one thing which causes a relative change in another thing.

protective reactor: A reactance coil used in a circuit to keep the current within a safe value when a short circuit occurs.

pull boxes: An iron box placed in a long conduit, or where a number of conduits make a sharp bend.

pull-offs: A hanger used to keep the trolley wire in proper place on a curve.

pulsating current: A current that flows in the same direction all the time, but rises and falls at regular intervals.

puncture: The breaking through insulation by a high voltage.

push button: A small contact device having a button which, when pressed, closes a circuit and causes a signal bell to ring.

push-button switch: A switch that opens and closes a circuit when a button is pushed.

push-pull transformer: A transformer used in radio work with a tap brought out at the center of the coil windings.

pyrometer: An instrument that indicates or measures temperatures higher than a thermometer will handle.

Q

Q: Abbreviation for "quantity" of electricity. The unit is coulomb or ampere-hours.

Q.S.T.: A radio code call—"Have you received the general call"?

quad: An abbreviation for quadruple telegraph; means Four.

quaded cable: A telephone or telegraph cable in which every two pairs (4 wires) are twisted together.

quadrature: Angle of 90 electrical degrees or quarter cycle difference between two alternating-current waves.

quarter phase: Same as two phase. The voltage waves are one-fourth of a cycle apart.

quick-break switch: A knife switch arranged so it will break the circuit quicker than when pulled open by hand.

R

R: Abbreviation for resistance, the unit of which is the ohm.

R.L.M.: Abbreviation for a dome type of lighting reflector.

r.p.m.: Abbreviation for revolutions per minute.

R.S.A : Railway Signal Association.

racing of motor: A rapid change or excessive speed of a motor.

raceways: Metal molding or conduit that has a thinner wall than standard rigid conduit used in exposed wiring.

racks and hooks: Supports for lead covered cables placed in underground manholes.

radial: In a straight line from the center outward.

radian: The angle at the center of a circle where the arc of circumstance is equal to the radius of the circle. It is 57.3 degrees.

radiation: The process of giving off or sending out light or heat waves.

radio: Referring to methods, materials, and equipment for communicating from one place to another without the use of wires between them.

radioactive: Giving off positive and negative charged particles.

rail bond: A short piece of wire or cable connecting the end of one rail to the next.

rating: The capacity or limit of load of an electrical machine expressed in horsepower, watts, volts, amperes, etc.

ratio: The relation of one number or value to another.

ratio arms: The two arms of a Wheatstone bridge whose resistances are known and form the ratio of the bridge.

ratio of a transformer: The relation of the number of turns in the primary winding to the secondary winding.

reactance: The influence or action of one turn of a coil or conductor upon another conductor which chokes or holds back an alternating current but allows a steady direct current to flow without any opposition.

reactance coil: A choke coil. It is used to hold back lightning and other high frequency currents in a circuit.

reactive current: That part of the current that does not do any useful work because it lags behind the voltage.

reactive load: A load, such as magnets, coils, or induction motors, where there is reactance which causes the current to lag behind the voltage.

reactor: Choke coils or condensers used in a circuit for protection or for changing the power factor.

reamer: A cone shaped tool used with a hand brace to remove the burr on the inner edge of conduit.

receiver: The part of the telephone that changes the talking current into sound that can be heard by the ear.

receiving sets: Devices used to receive radio messages and especially radio broadcast programs.

receptacle: A device placed in an outlet box to which the wires in the conduit are fastened, enabling quick electrical connection to be made by pushing an attachment plug into it.

receptacle plug: A device that enables quick electrical connection to be made between an appliance and a receptacle.

reciprocal: One divided by the number whose reciprocal is being obtained. The reciprocal of 2 is ½; of 3 is ⅓, etc.

recorder: A device that makes a record on paper of changing conditions in a circuit, apparatus, or equipment.

rectifier: A device that changes alternating current into continuous or direct current.

rectigon: Trade name for a battery charging rectifier.

red lead: Minimum, or peroxide of lead, used in making pasted battery plates.

re - entrant: Armature windings which return to a starting point, thus forming a closed circuit.

reflector: A device used to direct light to the proper place.

regenerative braking: Using electric motors on a car or locomotive as generators to slow down the train.

regulation: A change in one condition which causes a change in another condition or factor.

regulator: A device for controlling the current or voltage, or both, from a generator or through a circuit. Devices for controlling other machines.

relay: A device by which contacts in one circuit are operated by a change in conditions in the same or another circuit.

reluctance: The resistance to flow of magnetism through materials.

reluctivity: The reciprocal of permeability. The resistance to being magnetized.

remagnetizer: A large direct-current electromagnet used to magnetize the permanent magnets that have lost their magnetism.

remote control: Operating switches, motors, and devices located some distance from the control point by electrical circuits, relays, electromagnets, etc.

renewable fuse: An inclosed fuse so constructed that the fusing material can be replaced easily.

repeater: A device that reproduces the signals from one circuit to another.

repeating coil: An induction coil or transformer used in telephone work that has the same number of turns on each winding.

repulsion: The pushing of two magnets away from each other.

repulsion induction motor: An alternating current which operates as a repulsion motor during the starting period and as an induction motor at normal speed.

residual magnetism: The magnetism retained by the iron core of an electromagnet. Often the flow of current is stopped.

resistance: That property of a substance which causes it to oppose the flow of electricity through it.

resistance bridge: A Wheatstone bridge.

resistance furnace: A furnace where heat is obtained by electric current flowing through resistance coils.

resistor: Several resistances used for the operation control or protection of a circuit.

resonance: A condition in a circuit when the choke coil reactance is exactly balanced or equalized by a condenser.

resultant: The sum of two forces acting on a body.

retarding coil: A choke coil.

retentivity: Holding or retaining magnetism.

retriever: Device that pulls down the trolley pole of a car when the trolley wheel leaves the wire.

return circuit: The path the current takes in going from the apparatus back to the generator.

return feeders: Copper cables connected at different points of the rail to carry the current back to the generators.

reverse: Going in the opposite direction.

reverse current relay: A relay that operates when the current flows in the opposite direction to what it should.

reverse phase: A change in the phase of the current due to changing the generator or circuit wiring.

reverse power: Sending electric energy in the opposite direction in a circuit to the usual direction.

reversing switches: Switches used to change the direction of rotation of a motor.

rheostat: A resistance having means for adjusting its value.

ribbon conductor: A conductor made from a thin flat piece of metal.

right-hand rule: A rule used to determine the direction of flow of current in a dynamo.

ring armature: An armature with a core in the shape of a ring.

ring oiling: A system of oiling where a ring on the shaft carries oil to the top of the bearing.

ring system: Where two transmission lines from a station are joined together at a substation, thus forming a loop or ring.

risers: Wires or cables that are run vertically from one floor to another and supply electric current on these floors.

rocker arms: The arms to which the brush holders of a motor are fastened or supported.

rodding: Pushing short rods which are joined together through a conduit in order to pull a cable into it.

Roentgen rays: Similar to X-rays.

rosettes: A device to permit a drop cord to be attached to a ceiling outlet or fixture.

rotary converter: A direct-current motor with collector rings connected to the armature windings which changes alternating to direct current or the reverse; a synchronous converter.

rotary switch: A switch where the circuit is opened and closed by turning a knob or handle.

rotor: The part of an electrical machine that turns or rotates.

rotor slots: Openings punched in the disk of the rotor and in which the winding is placed.

r.p.m.: Abbreviation for revolutions per minute.

r.p.s.: Abbreviation for revolutions per second.

rubber-covered wire: Wires covered with an insulation of rubber.

rubber gloves: Insulated gloves worn by linemen when working on "Live" lines.

rubber tape: An adhesive elastic tape made from a rubber compound.

runner: The revolving part of a water turbine.

running torque: The turning power of a motor when it is running at rated speed.

runoff: The quantity of water flowing in a stream at any time.

S

s.: Abbreviation for second of time.

S.A.E.: Society of Automotive Engineers.

s.c.: Abbreviation for single contact.

s.c.c.: Abbreviation for single cotton-covered wire.

S.E.D.: Society for Electrical Development.

s.c.e.: Abbreviation for cotton enameled wire.

s.p.: Abbreviation for single pole.

S.S.: Abbreviation for steamship when placed before the name of the vessel.

s.s.c.: Abbreviation for single silk-covered wire.

S.K.F.: The trade name for a ball bearing.

safe carrying capacity: The maximum current a conductor will carry without overheating.

safety catch or fuse: A device that opens the circuit when it becomes too hot; often placed in base of appliances for heating liquids.

safety switch: A knife switch inclosed in a metal box and opened and closed by a handle on the outside.

salammoniac: Common name for ammonium chloride, NH_4Cl, used as electrolyte in primary cells.

salient poles: The ordinary poles formed at the end of a magnet as distinguished from consequent poles.

saturation curve: A curve showing the relation between the voltage produced by a generator and the ampere turns on the field coils.

Scott connection: A transformer connection for changing alternating current from two- to three-phase or the reverse.

seal: A piece of lead or metal used to close meter to prevent tempering.

second: $\frac{1}{60}$ part of a minute.

secondary: The circuit that receives power from another circuit, called the primary.

secondary battery: A storage battery.

secondary circuit: The wiring connected to the secondary terminals of a transformer, induction coil, etc.

secondary currents: Currents produced by induction due to changes in current values in another circuit.

section: An insulated length of line or circuit fed by a separate feeder.

sediment: Loose material that drops off storage battery plates and separators into bottom of cell.

segment: One of the parts into which an object is divided; often used to refer to commutator bars.

selector switch: A switch used in an automatic telephone system to locate an idle line.

selenium: A rare metal, the resistance of which changes when under action of light.

self-cooled transformer: A transformer in which the windings are cooled by contact with air or oil and without additional means for radiation.

self-discharge: The discharge of a cell due to leakage or short circuit inside of it.

self-excited: A generator in which the current in the field coils is produced by the generator itself.

self-induced current: An extra current produced in a circuit by change of the current flowing in that circuit.

self-inductance: The magnetic property of a circuit that tends to oppose a change of the current flowing through that circuit.

separators: Wood or rubber plates placed between the plates of a storage battery.

semaphore: A post or stand supporting a railroad signal.

separately-excited: A generator in which the current for the field coils is obtained from another generator or battery.

series: Connected one after another so the same current will flow through each one.

series arc lamp: An arc lamp in which the same current flows through all the lamps connected to the circuit.

series circuit: A circuit in which the same current flows through all the devices.

series generator: A constant-current generator used for operating a street lighting circuit where all lamps are connected in series.

series motor: A motor where all the current flows through the field coils and armature, because they are connected in series.

series-multiple: Same as series parallel.

series - parallel: An arrangement where several devices are connected into series groups and these groups are connected in parallel with each other.

series transformer: A current transformer. A transformer where the primary is connected in series with the circuit.

series winding: A wave-wound armature. A field coil winding through which the armature current flows.

service connections: The wiring from the distributing mains to a building.

service switch: The main switch which connects all the lamps or motors in a building to the service wires.

service entrance: The place where the service wires are run into a building.

service wires: The wires that connect the wiring in a building to the outside supply wires.

sheath: The outside covering which protects a wire or cable from injury.

shell transformer: A transformer with the iron core built around the coils.

shellac: A gum dissolved in alcohol, which forms a good insulating liquid.

sherardizing: Coating iron or steel with zinc to prevent rusting.

short: A contraction for short circuit.

short circuit: An accidental connection of low resistance joining two sides of a circuit, through which nearly all the current will flow.

short shunt: Connecting the shunt fields directly to the armature of a compound generator or motor instead of having them in parallel with armature and series fields.

short time rating: A device that can only operate for a short time without being allowed to cool.

shunt: A parallel circuit. A bypass circuit.

shunt coil: A coil connected in parallel with other devices and through which part of the current flows.

shunt field: A field winding connected in parallel with the armature.

shunt ratio: The ratio of current flowing through the shunt circuit to the total current.

shunt winding: A winding connected in parallel with the main winding.

shuttle armature: An H-type armature.

silicon bronze: A bronze or brass containing silicon and sodium which give it strength and toughness.

silicon steel: An alloy steel having low hysteresis and eddy current loss, used in transformer cores.

silk - covered wire: Small copper wires insulated by a covering of silk threads.

simplex circuit: A telegraph which sends in only one direction at a time.

simplex winding: A type of armature winding with two parallel paths from one brush to another.

sin: Abbreviation for sine of an angle; as sin 30°.

sine of an angle: In a right angle triangle it is the length of the side opposite the angle divided by the hypotenuse.

sine wave: The most perfect wave form. An alternating-current wave form.

single contact lamp: An automobile lamp which has one contact in end at base which makes contact with the side of the base and socket; the side of the base and socket completes the circuit.

single phase: A generator or circuit in which only one alternating-current voltage is produced.

single-phase circuit: A 2- or 3-wire circuit carrying a single-phase current.

single-phase motor: An alternating-current motor designed to operate from a single-phase circuit.

single-pole switch: A switch that opens and closes only one side of a circuit.

single-stroke bell: A bell that strikes only once when the circuit is opened or closed.

single-throw switch: A knife switch that can be closed to one set of contacts only instead of two, as with a double-throw switch.

single-wire circuit: A circuit using one wire for one side and ground for the other side or return conductor.

single re-entrant: An armature winding in which the circuit is traced through every conductor before it closes upon itself.

sinusoid: A sine curve.

six phase: A circuit or machine where the voltage waves are $\frac{1}{6}$ of a cycle behind each other.

skin effect: The action of alternating current that causes more of a current to flow near the outside than in the center of a wire.

slate: A rock that is cut into slabs and used for switchboards. It is a fair insulator.

sleet cutter: A device placed on the trolley wheel to cut or scrape sleet from the trolley wire of a railway system.

sleeve joint: Joining the ends of two wires or cables together by forcing the ends into a hollow sleeve and soldering them.

sleeving: A small woven cotton tube slipped over the ends of armature leads to give additional insulation.

slide wire bridge: A Wheatstone bridge in which the balance is obtained by moving a contact over a wire.

slip: The difference in speed between the speed of a rotating magnetic field and the rotor of an induction motor.

slip ring: A ring placed on a rotor, which conducts the current from the rotor to the external circuit. Collector ring.

slot: The groove in the armature core where the armature coils are placed.

slot insulation: Material placed in armature slot to insulate the coils for the core.

slow-burning insulation: An insulation that chars or burns without a flame or blaze.

smooth core: An armature where the conductors are bound on the surface instead of being placed in slots or grooves.

snap switch: A rotary switch where the contacts are operated quickly by a knob winding up a spring.

sneak current: A weak current that enters a telephone circuit by accident. It will not blow a fuse, but it will do damage if allowed to continue.

soaking charge: A low rate charge given to a storage battery for a long time to remove excess sulphate from the plates.

soapstone: A soft oily stone sometimes used for insulating barriers. The powder is used when pulling wires into conduit.

socket: A receptacle or device into which a lamp bulb is placed.

sodium chloride: Common ordinary salt.

soft-drawn wire: Wire that has been annealed and made soft; often being drawn to size.

soldering flux: A compound that dissolves the oxide from the surfaces being soldered.

soldering paste: A soldering flux prepared in the form of a paste.

solenoid: A coil of insulating wire wound in the form of a spring or on a spool.

solenoid core: The soft iron plunger or body placed inside a solenoid.

solid wire: A conductor of one piece instead of being composed of a number of smaller wires.

sounder: A telegraph relay that delivers a sound at the receiving end which the operator can understand.

south pole: The end of a magnet at which the lines of force enter.

space factor: The actual cross-sectional area of copper in a winding divided by the total space occupied by the insulation and winding.

spaghetti insulation: A closely woven cotton tube impregnated with an elastic varnish that is slipped over ends of bare wires to insulate them.

spark coil: An induction coil used to produce a high voltage which causes a spark to jump a gap.

spark gap: A device which allows a high voltage current to jump a gap.

spark plug: A threaded metal shell having a center insulated conductor, which is screwed into the cylinder of an automobile engine.

spark voltage: The lowest voltage that will force a spark between two conductors insulated from each other.

sparking at brushes: Small arcs or flashes occurring between the commutator and brush, due to poor contact or incorrect brush position.

sparkless commutation: Operation of a direct-current generator or motor without any sparking at the brushes.

specific gravity: The weight of any volume of liquid or solid divided by the weight of an equal volume of water; or of any gas divided by an equal volume of air.

specific resistance: The resistance of a cube of any material which is one centimeter long on each edge.

speed counter: An instrument that records the number of revolutions made by a shaft.

speed regulation: The per cent of full load speed that the speed of a motor changes, when the load is suddenly removed.

sphere gap: A spark gap formed between two spheres fastened to conductors.

spherical candle-power: The average candle-power from a light measured in all directions.

spider: A cast-iron frame with radially projecting arms on which the rotating part of an electrical machine is built.

splice: The joining of the ends of two wires or cables together.

splice box: An iron box in which cable connections and splices are made.

split knobs: Porcelain knobs made into two pieces to receive a wire or cable and held together by a screw.

split phase: Obtaining currents of different phases from a single-phase circuit by use of reactances of different value in parallel circuits.

split-phase motor: A three-phase motor that is operated by split-phase current obtained from a single-phase circuit.

split-pole converter: A synchronous converter with divided or additional field poles for regulating the voltage.

sponge lead: Porous lead used in the active material of the negative plate of an acid storage battery.

spot welding: Uniting two metals together by electric welding them at several spots.

square mil: The actual area of a wire or conductor expressed in mils. The $\frac{1}{1,000,000}$ part of a square inch.

squirrel cage: The arrangement of copper rods in cylindrical form and fastened to copper rings at each end of the rotor core of an induction motor.

squirted filament: The old method of forcing a soft material for a lamp filament through small holes.

staggering of brushes: Arranging the brushes on a commutator so they will not all bear or rub on the same place.

stalling torque: The twisting or turning power of a motor, just before the armature stops turning, due to heavy load being applied.

standard candle: A standard of lighting power.

standard cell: A primary cell that gives the legal standard of voltage.

standard ohm: The unit of resistance.

standard resistance: An accurate resistance that is used for comparison with unknown resistances.

stand-by battery: A storage battery connected to the distribution system to carry the load should the generators fail.

static machines: Generators that produce static electricity.

star connection: Connecting one end of each phase of a three-phase circuit or machine together, thus forming a common point called the neutral. A Y-connection.

starter: A device that enables a safe current to be supplied to a motor when starting.

starting battery: A storage battery designed to deliver current to a motor used for starting an automobile engine.

starting box: A rheostat used for a short time when starting a motor.

starting current: The current taken by a motor when starting.

starting motor: A motor used for cranking an automobile engine.

starting rheostat: A starting box.

starting torque: The turning power produced by a motor when the rotor begins to turn on that power required to start a machine at rest.

static charge: A quantity of electricity existing on the plates of a condenser.

static electricity: Electricity at rest as distinguished from electric current, which is electricity in motion.

static generator: A machine producing static electricity.

static transformer: An ordinary transformer in which all parts are stationary as distinguished from the earlier constant-current transformer with a moving coil.

stator: The stationary part of an induction motor on which the field windings are placed.

steady current: A direct current whose voltage does not change or vary.

step-down: Reducing from a higher to a lower value.

step-up: Increasing, or changing from a low to a higher value.

stop charge device: A device that disconnects a storage battery from the charging circuit when it is completely charged.

storage battery: A number of storage cells connected together to give the desired current and voltage and placed in one case.

storage cell: Two metal plates or sets of plates immersed in an electrolyte in which electric current

can be passed into the cell and changed again into chemical energy and then afterwards changed again into electrical energy.

strain insulator: An insulator placed in a guy wire to insulate it from the current-carrying wire.

stranded wires: Wires or cables composed of a number of smaller wires twisted or braided together.

stray current: Current induced in a conductor or core and which flows in these parts. The return current of an electric railway system that flows through adjacent pipes and wires instead of the regular return circuit.

stray field: Magnetic lines of force that do not pass through the regular path and therefore do not do any useful work.

stray flux: The lines of force of a stray magnetic field.

stray power: The power losses of an electrical machine due to heating effects, as friction, hysteresis, and eddy currents.

strength of current: The number of amperes flowing through the circuit.

strength of magnetism: The number of magnetic lines of force per unit of area.

strip fuse: A fuse made from a flat piece of metal.

Stubs' wire gauge: An iron wire gauge, often called Birmingham wire gauge.

sub-station: The building or place where one form of electrical energy is changed into another, as alternating current into direct current, high voltage to low, or the reverse.

sulphating: The forming of a hard white substance on the plates of a storage battery.

sulphuric acid: The kind of acid that is diluted and put in a lead storage battery.

superposed circuit: An additional circuit obtained from a circuit used for another purpose without interfering with the first circuit.

surface leakage: The leaking of current over the surface of an insulator from one metal terminal to another.

surges: An oscillating high voltage and current waves that travel over a transmission line after a disturbance.

surging discharge: A high voltage oscillating discharge.

susceptance: One of the components in an alternating circuit; the power component is called conductance and the wattless component is called susceptance.

susceptibility: The ratio of the amount of magnetism produced in a body to the magnetzing force.

suspension insulator: An insulator hung from a support and with the conductor fastened to the bottom of the insulator.

swinging cross: The blowing together of the wires of a transmission line, causing a short-circuit.

switch: A device for closing, opening, or changing the connections of a circuit.

switch blade: The movable part of a switch.

switchboard: The panel or supports upon w h i c h are placed the switches, rheostats, meters, etc., for the control of electrical machines and systems.

switchboard instruments: Meters mounted on a switchboard.

switch house or room: The part of the building in a power plant where the high voltage switches are located.

switch plate: A small plate placed on the plastered wall to cover a push button or tumbler switch.

switch tongue: The movable part of an electric railway track switch.

symbol: A letter, abbreviation, or sign that stands for a certain unit or thing.

synchronism: Alternating - current voltage waves that have the same frequency and reach their maximum value at the same instant.

synchronize: To bring to the same frequency and in phase.

synchronizer: A device for indicating when two machines are in synchronism.

synchronoscope: A n instrument which shows when two machines are in synchronism and which machine is leading the other in phase.

synchronous c o n d e n s e r: A synchronous motor operated without load and strong field current in order to improve the power factor.

synchronous converter: A direct-current motor fitted with collector rings and used to change alternating to direct current.

synchronous motor: An alternating-current motor whose speed is in proportion to the frequency of the supply current and the number of poles in the machine.

synchronous p h a s e advancer: A synchronous motor operated as a condenser to improve the power factor.

T

T: Abbreviation for temperature.

t: Abbreviation for time in seconds.

Ta: Chemical symbol for tantalum

T-connector: A connector joining a wire to two branch circuits.

T-splice: A connection joining the end of one wire to the middle of another one.

tachometer: An instrument that shows the number of revolutions per minute made by a shaft.

talc: Powdered soapstone.

tan: An abbreviation for tangent of an angle.

tangent: A straight line that just touches the circumference of a circle.

t a n g e n t galvanometer: A galvanometer operated by current passing through a coil overcoming the earth's magnetism.

tap: A wire connected some distance from the end of the main wire or conductor.

tape: A narrow strip of treated cloth.

tapering charge: Charging a storage battery at constant voltage. The rate of current flow will decrease as the battery becomes charged.

taping: Wrapping layers of tape around a wire, coil, or conductor.

teaser winding: An extra winding on the poles of a series wound dynamo.

teeth of armature: The projections between the slots in an armature.

telegraph: A system of sending messages by dot and dash signals

telegraph relay: A relay used in a telegraph circuit.

telegraph code: The dot and dash signals used for letters or words.

telephone: A device that transmits speech and sound from one place to another by electric currents.

telephone cable: A number of small insulated copper wires bound together and covered with paper, cotton, braid, or lead covering.

telephone condenser: A condenser used in a telephone circuit, made by rolling strips of tin foil between sheets of paraffin paper.

telephone cord: Several very flexible wires covered with a cotton braid. Used to connect one part to another.

telephone exchange: T h e p l a c e where all telephone lines end and connections are made from one line to another.

telephone jack: A receptacle into which a plug is placed when connecting one telephone line to another.

telephone receiver: A device that changes electric current in the telephone circuit into sound.

telephone repeating coil: A transformer used to reproduce the signals from one circuit to another.

telephone set: All the parts, such as transmitter, ringer, receiver, etc., installed for the subscriber's use on his premises.

temperature: Condition in regard to heat and cold.

temperature coefficient: The rate of change in resistance per degree change in temperature.

temperature correction: The amount that must be added to a reading taken at one temperature in order to make it comparable with the same reading taken at a standard temperature.

temperature rise: The difference in temperature between a certain part of a machine and the surrounding air.

tension: The degree of stretching; also sometimes used to refer to voltage, difference of potential, or dielectric stress.

terminal: A connecting d e v i c e placed at the end of a wire, appliance, machine, etc., to enable a connection to be made to it.

terminal lug: A lug soldered to the end of a cable so it can be bolted to another terminal.

terminal pressure: The voltage at the generator or source of supply.

Tesla coil: An induction coil on a transformer without an iron core, used to produce high frequency currents.

test clip: A spring clip fastened to the end of a wire used to make connections quickly when testing circuits or devices.

test lamp: An incandescent lamp bulb and socket connected in a circuit temporarily when making tests.

test point: The metallic end of an insulated conductor used in making tests.

test set: Electrical instruments and devices used for testing, mounted for convenient use.

testing transformer: A transformer designed to deliver a number of different voltages, and used in testing for defects.

theater dimmers: Variable rheostats connected in series with a lighting circuit to control the voltage to the lamps and amount of light produced by them.

thermal: Pertaining to heat.

thermocouple: Two different metals welded together and used for the purpose of producing thermo-electricity.

thermo-electricity: Electricity produced by the heating of metals.

thermo-galvanometer: A galvanometer operated by the heating effect of a current acting on a thermocouple.

thermometers: Instruments for indicating relative temperatures.

thermostat: A device that opens and closes a circuit when the temperature changes.

third-brush generator: A small generator placed on an automobile to charge a storage battery.

third-brush regulation: A generator whose voltage is regulated by armature reaction and the shunt field current obtained from a third brush bearing on the commutator.

third rail: An insulated rail, placed along side of the rails on an electric railway, which supplies the power to the cars.

three-phase: A generator or circuit delivering three voltages that are $\frac{1}{3}$ of a cycle apart in reaching their maximum value.

three-phase circuit: A circuit delivering three-phase current.

three-phase motor: An alternating-current motor that is operated from three-phase circuit.

three-pole: A switch that opens and closes three conductors or circuits at one time.

three-way switches: A switch with three terminals by which a circuit can be completed through any one of two paths.

three-wire circuit: A circuit using a neutral wire in which the voltage between outside wires is twice that between neutral and each side.

three-wire generator: A direct-current generator with a balancer coil connected to the armature windings and the middle point of the balancer coil connected to the neutral.

tie wire: A short length of wire used to fasten the overhead wires to a pin insulator.

time switch: A switch controlled by a clock that opens and closes a circuit at the desired time.

timer: A device that opens the primary circuit of an induction coil at the right time to produce a spark to fire the charge in an internal combustion engine.

tinfoil: Sheets of tin rolled out thinner than paper.

tinned wire: Wire covered with a coating of tin or solder.

torque: The twisting or turning effort.

torsion dynamometer: An instrument that measures the torque of a machine by twisting a calibrated spring.

track circuit: The circuit through the rails and bonds.

track return: The return circuit formed by the rails and bonds of a track.

train lighting battery: A storage battery used to furnish electricity for lighting railroad cars.

transformer: A device used to change alternating current from one voltage to another. It consists of two electrical circuits joined together by a magnetic circuit formed in an iron core.

transformer coil: A part or one of the windings of a transformer.

transformer efficiency: The power delivered by a transformer divided by the power input to it.

transformer loss. The difference between the power input and output.

transformer oil: Oil used in a transformer to insulate the windings and carry away the heat.

transformer ratio: The ratio of the primary to the secondary voltages.

transformer substation: A substation where the alternating-current voltage is stepped up or down by use of transformers.

transite: A kind of asbestos lumber used for insulating barriers in dry places.

transmission line: High voltage conductors used to carry electrical power from one place to another.

transmitter: The telephone device that receives the speech and changes it into electric current.

transposition: Changing the relation of telephone and electric light wires to each other in order to equalize the inductance and prevent cross talk.

trickle charge: A low rate of charge given a storage battery.

triphase: Same as three-phase.

triple-pole switch: Same as a three-pole switch.

trolley wire: A wire supported over the tracks of an electric railway which carries the power for operating the cars.

true resistance: Actual resistance measured in ohms as compared to counter-electromotive force.

trunk: The wires or circuits between switchboards or telephone exchanges.

tube insulator: Insulating material made in the form of a tube and used to carry conductors through walls and partitions.

tumbler switch: A switch similar to a flush push button, but operated by pushing up or down on a short lever.

tungar rectifier: A rectifier using a tungar bulb made or licensed by the General Electric Company.

tungsten: A very hard metal with a high melting point that resists the effects of arcing.

tungsten filament: A filament made from tungsten and used in a lamp bulb.

tungsten steel: An alloy of steel and tungsten which produces a hard tempered steel which retains this property when heated a dull red.

twin cable: Two insulated wires running side by side without being twisted and covered with a braid.

twisted pair: Two rubber-covered telephone wires twisted together and used to connect subscriber's set to overhead wires or cable.

two-phase circuit: A circuit in which there are two voltages differing by one quarter of a cycle.

two-phase motor: A motor made to be operated from a two-phase circuit.

two-phase generator: A generator producing two-phase current.

two-pole: A switch that opens or closes both sides of a circuit or two circuits at one time.

two-wire circuit: A circuit using two wires.

U

ultra violet rays: Light rays that are beyond the violet color and not visible.

unbalanced load: A distribution system where there is a greater load on one phase or side than on the other.

undamped waves: Radio waves whose maximum rise and frequency is constant.

under-charged battery: A storage battery that has not been sufficiently charged.

under-compounded: A compound-wound generator in which the voltage drops as the load increases.

under-cut mica: Cutting the mica between commutator segments below the surface so it will clear the brush.

underground cable: A cable insulated to withstand water and electrolysis and placed in underground conduit.

underload circuit breaker: A conduit breaker that opens when the load drops below a certain value.

underload relay: A relay that operates another circuit when the load drops below a certain value.

Underwriters' Code: The National Electric Code.

unidirectional current: Current that flows in one direction.

uniphase: A single-phase alternating current.

unipolar: Having one pole.

unit price: Cost of one piece, foot, pound, or whatever number is taken as a unit for that particular material.

unloader: A device that removes the load from a machine, such as a compresser, when a motor is starting it.

V

V: Abbreviation for volts or potential difference.

V.T.: Abbreviation for vacuum tube or electron tube.

vacuum cleaner: A machine that sucks dust and dirt out of rugs, drapes, upholstery, etc.

vacuum impregnated: Filling the spaces between electric parts with an insulating compound while they are placed in a vacuum.

vacuum tube: Any kind of a bulb or tube from which the air has been removed.

vapor: A gas from a substance that is ordinarily a liquid or solid.

vapor rectifier: A mercury arc rectifier.

variable condenser: A condenser whose capacity can be varied.

variable resistance: A resistance that can be changed or adjusted to different values.

variable-speed generator: A generator operated at different speeds with a method of regulation which causes it to deliver a constant voltage.

variable-speed motor: A motor whose speed depends upon the load.

Varley loop: A method of locating a cross, short-circuit, or ground on telephone or telegraph lines.

varnished cambric or cloth: Cotton cloth treated with an insulating varnish.

vector: A line whose length and direction represents a certain physical quantity.

vector diagram: A diagram that shows relations by use of vectors.

verdigris: A substance called copper sulphate that forms on copper by the action of sulphuric acid.

vibrating rectifier: A device that changes alternating current into direct current by means of a vibrating contact that closes the circuit for one-half of the cycle and opens it when the flow of alternating current is in the opposite direction.

vibrator coil: An induction coil used as an ignition coil.

volt: A unit of electrical pressure or electromotive force.

voltage coil: A coil connected across the line so that the current flowing through it changes as the voltage changes.

voltage drop: The difference in pressure between two points in a circuit caused by the resistance opposing the flow of current.

voltage loss: The voltage drop.

voltage regulator: A device for keeping a constant voltage at a certain point.

voltaic battery: A number of primary cells connected in series or parallel.

voltammeter: A voltmeter and ammeter combined in one case and using the same movement, but having separate terminals.

volt-ampere: The unit of apparent power; it is the product of the pressure times the current.

voltmeter: An instrument that shows the pressure or voltage of a circuit

vulcabeston: An asbestos and rubber composition used to make moulded parts.

vulcanite: A kind of hard rubber.

vulcanized fiber: An insulating material made of paper and cellulose under heavy pressure.

W

W: Abbreviation for watt.

W.A.E.I.: Western Association of Electrical Inspectors.

wall box: A metal box for switches, fuses, etc., placed in the wall.

wall insulator: An insulating tube used to protect a conductor passing through a wall.

wall socket: An electric outlet placed in the wall so that conductors can be connected to it by means of a plug.

water-cooled transformer: A large transformer having coiled pipes inside it through which water passes.

water rheostat: A rheostat that has its terminals placed in water through which the current flows.

watt: The unit of electric power.

watt-hour: The use of a watt of power for an hour.

watt-hour meter: An instrument that records the power used in watt-hours.

watt meter: An instrument used to indicate the power being used in a circuit.

watt minute: A power of one watt being used for one minute. $\frac{1}{60}$ of a watt-hour.

wattless: Not having any power or doing any useful work.

wave meter: An instrument used to determine the wave length or frequency of a radio broadcasting station.

wave winding: An armature winding with the end of the coils connected to commutator bars that are nearly opposite each other in a 4-pole machine.

weatherproof: Constructed so it will resist the action of rain, sun, etc.

welding transformer: A transformer built to deliver a large current used to heat metals to a welding temperature.

welding flux: A material, usually borax, used to remove scale from the joints being welded.

Western Union splice: A method of uniting two wires together by wrapping each one about the other.

Weston cell: A primary cell that has a constant voltage and used as a standard source of electrical pressure.

wet storage: A method of keeping a storage battery when it is not being used without removing the acid or plates.

Wheatstone bridge: An electrical balance used to measure resistance by comparing a known resistance with an unknown.

windage: The resistance of air against the rotating part of a machine.

wiping contact: A contact that rubs between two other contacts.

wire: A slender rod of drawn metal.

wire gauge: A method of expressing the diameter of different wires.

wired radio: Transmitting radio messages along telephone, electric light, and power lines instead of directly through the air.

wiring connector: A device for joining wire to another.

wiring symbols: Small signs placed on a wiring diagram to indicate different devices and connections.

wood separator: A thin sheet of wood placed between the plates of a storage.

wrought iron: A kind of iron that can be easily magnetized.

X

x: A symbol used to represent an unknown quantity.

x: A symbol for reactance, expressed in ohms.

X-ray: A kind of ray that passes through most materials as if they were transparent.

Y

Y: A symbol for admittance; the unit of which is mho.

Y-connection: A star connection; the joining together of one end of each phase of a 3-phase machine.

yoke: The iron frame of a generator or motor to which the magnetic pole pieces are fastened.

Z

Z: Symbol for impedance.

zero potential: Not having any voltage or pressure.

zinc battery: A primary cell in which the electric current is produced by zinc plates immersed in an electrolyte.

INDEX

A

Air conditioning, 167-168, 231
Airstat, 191
Ampere, 10
Anchor strip, wooden, 115
Anchoring devices, 135-138
Appliances, large, 152-158
Apprentice, 247
Aquastat, 190-191
Architect's scales
   flat, 3
   triangular, 4
Arcing, 41, 45
Auger bits
   choosing, 110
   sharpening, 110-111

B

Bell system
   batteries, 72, 74
   chime, 74-75
   push, 72-73
   return-call, 72-73
   simple circuit, 72
   transformer, 74
Bonding jumper, 41
Boring holes
   by auger bits, 110-111
   in joists, 109
Boxes, outlet, 89-91
Brace, hand, 140
Branch circuit
   estimating, 235-240
Bushings, 129, 131

C

Cable stripper, 97
Cable wiring
   flexible armored, 98-101
      applications of, 99
      preparation for, 100-101
      requirements for, 99
      types of, 98-99
   for branch circuit cable, 102
   for service entrance, 101-102
   for underground feeder, 102
   in service extensions, 104
   in underplaster extensions, 102-104
   nonmetallic sheathed, 94-98
      applications of, 95
      methods of wiring new buildings,
         96-98
      requirements for, 95
      types of, 94

Calking tools, 135
Ceiling finish, 107
Ceiling height, 106
Ceiling outlet box, mounting, 107
Chases, 182-183
Chime, 74-75
Chisels, 140
Circuit
   parallel, 16-18
   series, 15-16
Circuit elements, 10
Cleat, 86
Concrete walls,
   installing outlets in, 181-182
Conductor, 18
   identified, 42, 72
   neutral, 21
   size, 19
   unidentified, 72
Conduit
   bushings and locknuts, 129
   couplings, 129
   E.M.T. type, 127
   elbows, 128-129
   flexible type, 127
   galvanized, 43
   methods in wiring, 142-151, 179
   rigid type, 126, 176
   thin-wall, 129
   tools for work, 138-142
   wire capacity of, 138
Conduit wiring
   anchoring devices for, 135-138
   fastening devices for, 134
   fittings, 129-131
   installing outlet boxes in, 131-134
   materials, types of, 126-127
   tools and methods, 138-151
   wire capacity of, 138
Connector
   for flexible armored cable, 100
   Scotchlok, 28
   Sherman fixture, 28
   solderless, crimp type, 28
Connectors, conduit
   for flexible type, 131
   for thin-wall tubing, 130
   used with adapters, 131
Construction procedures
   for multi-family dwellings, 175-183
Contractor's equipment, 140-142

Control equipment, automatic
 airstat, 191
 aquastat, 190-191
 damper, 193-194
 draft control, 192
 furnacestat, 191
 heat, 194
 pressurestat, 191
Cooking top, 229
Cooking units, built-in, 156-157
Costs, *see* Estimating
Couplings
 conduit, 129
 indenter type, 130
 self-locking, 130
 setscrew type, 130
 watertight, 129
Current
 basic concepts of, 10
 single-phase, 20
 three-phase, 20
Cutting a pocket, 121-122

D

Damper, 193-194
Demand load, 218, 220-224, 229
Dies and die stocks, 141-142
Dishwasher, 229-230
Draft control, 192
Dryer
 clothes, electric, 157-158, 228-231
Dwellings, multi-family
 construction procedures for, 175-183
 emergency lighting for, 189
 service considerations for, 183-188
 wiring methods for, 174-175

E

E.M.T., 127
Elbow, conduit
 former, 128
 thin-wall, 128
 types of, 128
 use of, 128
Electrical concepts, fundamental, 9-10
Electrical connections
 importance of, 22
 making splices, 25-32
 preparing, 22
 removing insulation, 23
 soldering, 23
Emergency lighting
 for multi-family dwellings, 189
Estimating
 branch circuit and fixture schedules
  for, 235-236
 branch circuit material schedules
  for, 236-240
 definition of, 233
 final considerations in, 246

form for, 243-245
 labor unit schedule for, 241-243
 other methods of take-off in, 240
 preliminary steps in, 234-235
 service and feeder material
  schedule for, 240-241
 short-cut method for, 246-247
Expansion shells, 135
 Ackerman-Johnson shell, 135
 calking tool, 135
 Rawl anchor, 135-136
 Rawl drives, 136
 Rawl plugs, 135
Extension rings, 133-134
Extensions
 surface, 104
 underplaster, 102-104

F

Farm wiring, 218-224
Fastening devices, 134
Fishing methods, 117-120
Fittings, thin-wall, 129
Floor construction, 108
Floor finish, 107
Floors
 concrete, 176
 wooden, 123
Form, estimating, 243-245
Furnacestat, 191

G

Garbage disposal unit, 229-231
Grounding
 armored cablework, 48-49
 bonding conduit, 47-48
 bushing, 46-48
 conductor, 41-42
 conduit system, 45
 danger, 41, 45
 electrode, 40-41
 equipment, 40, 43
 inadequate, 44-45
 nonmetallic sheathed cable, 49
 rods, 42-43
 soil conditions, 42-43
 system, 40-41
Grout, 183

H

Heat
 control equipment, 194
 pump, 167
Heater, infra-red, 163
Heater, resistance, 163
Heater, unit, 167
Heater, water, 158
Heating, space ·
 baseboard, 160
 ceiling, 162-163
 central, 158-159, 230

duct, 159
floor furnace, 160
practical considerations for, 168
wall, 160
Heating system control
gas burner, 198-199
oil burner, 195-198
stoker, 195-196
Hickeys, 142
Hinge point, 62-63
Hog house, 220

I

Identified conductor, 42, 72
Indenting tools, 140
Induction, 57-58
Infra-red heater, 163
Insulation
for open knob-and-tube type,
81-82, 87
for outside wiring, 217
stripping, 23
wire, 22
Intercom
combination, 72
four-station, 76-78
two-station, 75

J

Journeyman, 247
Jumper, bonding, 41

K

Kilowatt, 13
Knockouts
concentric, 47-48, 131
cutter, 140
Knob-and-tube wiring
advantages of, 80
concealed type, 88-91
method of wiring new buildings,
91-92
method of wiring old buildings, 92
open type, 81
precautions against dampness, 86-87
protection of, 83-85
separation of conductors, 82
stringing, 87-88
support for, 85
surface clearance, 82
types of, 80-81

L

Labor, estimating, 241-247
Lighting fixtures, special
fluorescent lamp, 124-125
recessed, 123-124
Locknuts, 129, 131
Loom, 84
Loud-speaker/intercom, 78
Lugs, solderless, 30

M

Meter shunt, 41

Methods used in conduit work
fishing, 149-151
to install thin-wall type, 148-149
to make offsets in rigid type, 145-147
to make offsets in thin-wall type,
147-148
to make right-angle bend in rigid
type, 142-145
to make right-angle bend in thin-wall
type, 147
Motors, wiring for, 168-173

N

National Board of Fire Underwriters, 2
National Electrical Code
general description of, 2
in Canada, 2
requirements of
for appliances, 157
for cable wiring, 99, 101-104
for conduit wiring, 129
for emergency lighting, 188
for estimating, 234, 238
for farm wiring, 218-222
for grounding, 41-47
for heating, 162
for identified conductor, 72
for knob-and-tube wiring, 82, 87-88
for motors, 169-173
for multi-family dwellings, 177,
183-184, 186, 188
for outside work, 215-218
for residential design, 153, 177,
226-231
for service switch, 39
for special utility circuits, 7
for wire capacity of conduit, 138
for wiring old residences, 111
National Fire Protection Association, 2
Network system, 20-21

O

Offsets, 145-147
Ohm, definition of, 10
Ohm's law, 10-13
Outlet boxes, 107, 131-134
Outlets
installing in concrete walls, 181-182
installing switch, 114-116
locating, 112-114, 178
plug receptacle, 116
Oven, electric, 229
Overcurrent protection, 39-40

P

Panelboards, 186-187
Pigtail splice, 25
Pipe benders, 142
Plan, typical, 4-7
Plan, yard distribution, 221-224
Plumb-bob, 140
Pocket, cutting a, 121-122

Polarity wiring, 55
  in conduit or cable, 100
  reasons for, 56
Pole line, 216
Pole riser, 35, 36
Poultry house, 219-224
Pour job, 176
Power
  definition of, 13
  equation, 14
  loss, 19
Pressurestat, 191
Profit, 246-247
Protector relay, 195
Pump, heat, 167
Pump house, 221-224

R

Range, electrical, 153-157, 229-231
Rawl
  anchor, 135-136
  drives, 136
  plugs, 135
Reamer, 140
Regulations, purpose for, 1-2
Relay, protector, 195
Remote control
  advantages of, 204-205
  equipment for, 205-208
  how it operates, 201
  planning, 208-211
  procedure for wiring, 211-214
  purpose of, 201
  relay, 206-209, 211-212
  switches, 207-208, 210-213
  transformer, 205, 210
  use of in wiring systems, 203
  wire, 206
Removing
  baseboard, 121
  floor boards, 120-121
  trim, 120
Resistance formula, 18
Resistance heater, 163
Residential designs, 109
Residential wiring
  for multi-family dwellings, 183-188
  for old houses, 111-125
Rule, folding, 140

S

Saddles, 146-147
Saw, keyhole, 140
Scotchlok connector, 28
Scuttle hole, 116-117
Section drawing, 106
Service and feeder, estimating, 240
Service disconnect switch, 229
Service drop, 184, 217
Service switch
  requirements of, 39
  type of, 39
Service wiring
  grounding, 40-45
  identified conductor, 42
  overhead connections, 37
  service drop, 33-35
  switch, 35, 39
  underground connections, 35
  watt-hour meter, 49-54
  wires, 35
Sheetrock, 115-116
Sherman fixture connector, 28
Skinning, 22
Slab
  definition of, 176
  ground, 176
  pan, 180-181
  upper floor, 177
  work, 176
Soldering
  copper wire, 23-24
  paste, 24
Solderless
  connector, crimp type, 28
  lugs, 30
  splices, copper wire, 28-30
  terminals, 30
Specifications, 233
Splices, aluminum conductor, 31-32
Splices, copper wire
  common types of, 25
  solderless, 28-30
  taping, 26-28
Stacking, 187-188
Story height, 106
Stripper, cable, 97
Sweeps, 128-129
Switch
  box cover, 133
  double-pole, 61
  electrolier, 70-71
  four-way, 67
  leg, 58-59
  loop, 58-59
  mounting in knob-and-tube type, 87
  pilot-light, 71
  rotary, 63
  service, 35, 39
  single pole, 58-59
  symbol, 5
  three-gang, 60-61
  three-way, 61
  two-gang, 60
Switch box, mounting, 115
Switchboards, 186-187
Symbols, electrical
  bell pushes, 7
  fuse panel, 7
  lights, 7

outlets, 5
switches, 5

T

Tape, steel, 140
Tapes
  plastic, 27-28
  rubber and friction, 26
Tee splice, 26
Terminals, solderless, 30
Testing
  circuit, 111-112
  set, 111
Thermostat, 158, 160, 164-167
Timeswitch, 187
Toggle bolts, 134
Tools
  caulking, 135
  miscellaneous shop, 140-142
  small, 138-140
Transformer, 40, 44, 74
Travelers, 62

U

Underground connections, service
  wiring, 35
Underground feeder, cable wiring
  for, 102
Underplaster extension, cable wiring
  in, 102-104
Underwriters Laboratories, Inc., 2, 223
Unidentified wire, 72

V

Vapor-proof fixture, 218-219
Vise, pipe, 140
Volt, definition of, 10
Voltage drop, 18-19

W

Wage scale, 247
Wall obstructions, passing, 118-120
Water heater, 158
Watt, definition of, 13
Western Union splice, 25
Watt-hour meter
  dial markings, 50-51
  installation, 53-54
  internal diagram of, 52
  meter constant for, 51
  reading a meter, 51
  three-wire, 52
  use of, 49-50
  wiring, 52
Wire
  color coding, 55-56
  hot, 55
  insulation, 22-23, 41, 81-82, 87, 217
  lead-in, 33
  main or phase, 20
  neutral, 21, 40, 55-56, 72
  service, 35
  size, 18-19
  stripper, 23
Wire nut, 28-29
  Scotchlok, 28
  Sherman fixture, 28
Wiring, outside, 215-218
Wiring methods
  for multi-family dwellings, 174-175
  for older houses, 111-125
Wrenches, 140

Y

Yard pole, 222, 224